I0049033

Alien Physics

How Nature and the Universe Really Work

Transcribed by Peter A Jackson FRAS

January 2024

2nd Edition. Revised, plus E Book, August 2024

Alien Publishing Inc. c/o Orchard House CT5 3EX. UK.

Figures.

Acknowledgements

To my mother who taught me wisdom, my flight engineer father for his practical skills and introduction to sailing. His RAF contemporary Freeman Dyson, the Cambridge mathematician, physicist and my main mentor who's unconstrained yet rigorous approach to revealing new understandings of nature greatly inspired me. To John Minkowski, optics specialist and co-author of key papers. To my wife Judith, from a family of yet more Cambridge mathematicians and Physical Chemist Professor and Dean. David Whiffen. Judith completed her own PhD with meagre support from me while greatly helping me in transcribing and proof reading this book. Lastly, and mostly, to those who developed and/or communicated the new physics herein solely in order to help in the advancement of mankind's intellectual development and his understanding of nature and the universe, whoever and wherever they may be or might have been.

About the Author

Peter Jackson set school exam records and studied a wide range of subjects at 3 institutions over 15 years, including Physics specialisms, astronomy, philosophy maths, logic, cognitive sciences, architecture, and engineering. He gained a reputation for innovative thinking and problem solving inspired by a principal mentor Freeman Dyson. He was soon given the lead consultant role in delivering the UK's largest petro-chemical project then also the UK's largest windfarm and many other green & renewable energy projects.

Peter is a fellow of the Royal Astronomical Society. His research in Physics, fluid dynamics, cosmology and space architecture was continued alongside successful consultancy practice. Links to academia continued with student mentoring work at two universities until after early retirement to concentrate on research.

Experimentation included repeating Hafele & Keating's trip round the world with an atomic clock, confirming their critical data consistent with Einsteins revised 1952 interpretation of Special Relativity, still not yet widely understood or adopted but explained in Peter's papers.

Married with three children, Peter competed in various sports, becoming a national champion and representing team GB as a yacht racing skipper & helmsman. He still races the successful 42ft racing yacht he co-designed.

His fundamental physics contest essays have had an unprecedented set of top peer scores, including top peer score in the 2015 maths essay including a new analysis of the cause of paradoxes besetting most logic and the consequential issues with maths. The essays never feared to use more logical departures from older doctrine so never gained official recognition or awards.

Peter has however been published widely including a solution to the Babcock & Bergman emission anomaly, on Einstein's final updated interpretation of the Theory of Special Relativity, on a new Galaxy Evolution model, Gravity, and a quasi 'causal' measurement sequence for Quantum Mechanics. A *cyclic* galaxy evolution model was presented to the AAS 2022 conference and was well received, shown to also suggest a Cyclic Cosmology able to resolve the pre 'Big Bang' state problem.

His 2022 Springer-Nature 'Foundations of Physics' paper with John Minkowski identifying a causal 'Measurement Problem' interaction sequence proved able to reproduce the data from the QM 'beam splitter' type experiments. Lack of such explanation led to the 'unphysical' theory of QM, to which Einstein (including in 'EPR') objected. Peter's 'Discrete Field' model (DFM) identifies how that solution to QM 'weirdness' enables heuristic unification with Special Relativity and leads to a 'Quantum Gravity'.

Publications Archive;
https://independent.academia.edu/JacksonPeter/Papers

Table of Contents

1. Preface

It should be clear to most by now that man's understanding of Nature and the Universe is flawed or incomplete, so some fundamental 'key' to the truth seems to be missing. We can reveal that it is. A logical assumption may be that the physics of more advanced civilizations is rather 'alien' to most human doctrinal assumptions. Indeed we shouldn't assume that all mankind's 'Laws of Physics' are complete & apply to the whole universe. Belief that they *should* do so may have slowed advancement of our theoretical understanding. This book passes on the fundamentals of the key elements used widely by greater intelligences.

The likely requirements for a 'translator' of alien concepts are receptiveness, good visualization skills and the ability to express them in the writing. An understanding of natural resistance to change, and how that can slow intellectual evolution, is likely also helpful. Humanity seems to appear to the aliens to be a unique and valuable civilization, but also one where technological advancement has advanced faster than theoretical comprehension and intellect, to an extent that poses dangers to its own survival.

Elsewhere, discussion seems to be ongoing about whether a 'tipping point' has been reached, needing more intervention. Earlier interventions apparently proved to be of limited value, so it appears to be recognised that care is needed. There also seems to be a 'prime directive' as Star Trek's Jim Kirk applied; as a *'non-interference'* rule. But nothing is absolute. It seems that no direct contacts are thought to be appropriate or desirable. During development of Earth's 'physics' many largely correct concepts have been proposed but not adopted. Their existence allows avoidance of giving direct advice as opposed to just carefully identifying and 'feeding back' such useful concepts, corrected as required.

The success of this project is recognised as far from guaranteed. Each human sees nature differently, so many and widely varying proposals of 'new physics' exist both within and without academia. Some fall into the 'nearly correct' category, but all have the effect of *closing* our minds to *other* new concepts. Skepticism is good but can cut both ways. Valuable for dismissing invalid ideas, but also damaging advancement by dismissing unfamiliar concepts. Revealing that this book contains 'Alien Physics' shouldn't cause panic because most will of course dismiss or ignore the likelihood of an alien source.

The veracity of the physics herein should be self-evident when properly analysed. Exactly where it 'comes from' isn't known or of importance. All knowledge is ultimately from 'the stars'. Translators need not know the source of content, just express the explanations and rationales given with the best of intentions, so please don't shoot the messenger!

The content doesn't pretend to be a *'theory of everything'* (TOE) or *'Grand Unification'* theory (GUT), both likely fallacious concepts. Wolfgang Pauli was right responding to a comment he'd *'got there'* with a cartoon artwork of a few lines and squiggles captioned; "*I can paint like Titian: only a few details are missing*". Pauli's *'exclusion'* was a correct concept. Alien physics uses and extends it to all dynamic/kinetic states k, k', k" etc. and provides clarifications, but it still leaves much detailed work to be done.

To *'condition'* readers, a first introduction to wider thinking may be a dimensional 'Mind spin'. Going beyond our familiar 'scale range'; subatomic to universal, first by expanding, past the galaxy as a vortex, then the universe as another, then as one of many particles, then keep going into a far larger world, however you wish to imagine it. Then, as if on elastic, shrink again to Earth scale, but keep going downwards, until particles become universes, then smaller ones as galaxies, then onto a planet, again whatever you wish to imagine but with living creatures. After a short period, you expand back 'upwards', through each scale and back past Earth to the first new world, but then onwards through that galaxy and universes to arrive at yet another, larger still, with a whole new civilization.

You may interact at any stage you wish, but then you shrink once again and go down past the first, past Earth and past the smaller world to its particles as universes then galaxies, and on to yet another. The process continues until your memory struggles to recall all the worlds and their order. Then slow down, visiting each again, finally settling back to Earth. After a few trips, often when lying half asleep a wider view of the universe emerges even without drugs! No point to these journeys was immediately clear, but deeper exploration of our minds potential does seem beneficial, so do give it a go!

Back to physics, the biggest *departure* from the correct route in decoding nature is identified as the switch to almost solely relying on 'mathematics' to advance understanding. False assumptions led to comprehension failing in key areas, it was easier to *'shut up and calculate,'* so abandon attempts to visualise and rationalise, damaging our intellect. 'Language' contributed to the problem. Imagine yourself studying the language of another species & find there are 7,100 *different* languages in use! 'English' is used as the common and definitive *'language of physics'* but English *excludes* 85% of humankind, unlike mathematics. A better and more universally understood written language may help, or extend English. Videos & VR will help dynamic 3D visualization, but concise linguistic descriptions are also needed for now.

English is a useful language when all its nuances are understood so is used here. The excluded 85% helped maths to become the *global* 'language of physics', but has now seriously degraded man's skills of visualization and dynamic rationalization.

Another problem is inherent in Boolean / binary mathematical modelling. Based on an early flawed concept, it can only ever an 'approximate' nature. Some have seen that the 'higher orders' aren't yet rationalised. The 'error' was retaining the Greek *'indivisible atom'* once atoms were 'divided'. The old concept was fine but the 'atom' came NOT to mean the 'smallest possible' entity.

Suggestions that maths is somehow 'wrong' and can 'mislead understanding' will doubtless shock many, but precisely *what* parts are wrong needs to be understood. The resulting 'mathematical fallacies' and paradoxes (mostly identified) are removed, with a better Physical 'Law' replacing the '*Law of the excluded*' middle between 0 and 1 with a sine curve distribution; '*Law of the **reducing** middle*'. Alien physics utilises our fractal higher orders, which

5

means that the Gaussian and Bayesian non-linear distribution approaches are valid. Nature isn't then well described by integer or binary maths.

The other main topics covered here include the main 'pillars' of current physics, Einstein's Special Theory of Relativity (**SR**) and Quantum Mechanics (**QM**) as well as Gravity and Cosmology. Each re-emerges as consistent with the others, causally and also then free of paradox. Most other areas or 'divisions' follow, but are no longer divided, being both 'reductionist and 'holistic' at the same time. The *reason* the speed of light 'c' is the value it is also logically and coherently emerges along with the apparent 'superluminal' speeds found!

Mankind's objective 'Scientific Method' (**SM**) would be commendable if only it were consistently used! All should be aware that humankind still relies significantly on 'belief' systems instead, resisting challenges to embedded foundational *beliefs which are really only flawed assumptions.* Alien Physics will challenge those prior assumptions and beliefs, but you'll find the current random pile of 'jigsaw puzzle' pieces will then perfectly slot together. The picture they reveal may be initially unfamiliar but will be found correct and simpler once embedded. Alien Physics will then *look* 'alien' to all with embedded beliefs (i.e. all!) but will prove *not* to be when properly tested using the SM.

Remember that good physics involves making and rigorously testing hypotheses, however unfamiliar they may first appear. No problems can be solved; "*using the same kind of thinking that created them.*" (Einstein). Only new 'novel' physics will advance understanding, by definition, but only a few of the key aspects are genuinely 'new'. The first corrects a key paradigm over 100 years old, the bounded *fluid* medium Einstein argued must be retained since 1919 and 1921 (Leiden lecture), but **not** the old *'Aether'* with its fatal *immobility* and so 'absolute' kinetic state Fluid media have 'flows, so the *shear planes* bounding the inertial systems k, k', k" etc.

There is close equivalence between real smaller, finite and 'bounded' kinetic states k, (which need higher order maths to model), and the *'curled up'* dimension of String & 'Brane' theories. String theory was born from Euler's equations, but has not yet produced, or 'tied up', any consistent or physical description of nature. However, readers will find such a description can now coherently emerge.

Short Bibliography of abbreviations.
SR & GR. Einstein's Special & General Relativity.
QM. Quantum Mechanics, EM. Electromagnetism.
IP/ IS/ IGM. Inter-Planetary/Stellar/Galactic Media,
DFM. 'Discrete Field Model.' As Einstein's '52 SR.

DART. Double Asteroid Redirection Test.
EPR Paradox. Einstein Podolsky, Rosen Paradox.
IRF. Inertial Reference Frame (or kinetic system).
ECI/ECRF. Earth's Orbiting & Rotating rest frames.
LT. Lorentz Transformation. (Gamma).
FSC. Fine structure Constant 'Alpha'.
TZ. 2-Fluid Plasma Transition Zone (As Maxwell's 'Near/Far field' transitions).

2. Sub-matter scale Fluid 'Condensate'.

The smallest form of 'condensed' matter are the free fermion e+/- 'pairs', known as 'electrons' and 'positrons' (+ 'quarks'), or combined as 'Majorana Fermion' dipoles, recognizing the fact that 'reversal' or 'spin flip' can reveal the two as simply the two hemispheres of dipoles! Later we'll see how these form as vortex pairs each side of condensate 'shear planes', as universally found in all fluids and gases.

It's been consistently ignored by humankind that these physical entities must be 'constituted' of some 'substance'. Some seem content to assume they 'pop up' from an 'empty vacuum' by magic! They can't and don't of course, as explained below.

So here's the first shock. There's no 'magic' in alien physics! The condensate is merely a smaller scale 'granularity', indeed just the 'next phase down' from the 'condensed matter' it creates by forming larger scale vortices, or (toroidal) vortex pairs, so forming free fermion 'space plasma'.

The condensate shares most characteristics with larger scale 'gasses'. It is "what is waving" as the EM fluctuations we find coupling with the fermions. Here we employ the familiar word 'electrons' to represent the fermions, the smallest 'size' with the ability to *couple with* (absorb & re-emit) EM wave fluctuations to allow the detectors to be 'triggered' and so produce new states we call 'measurements'.

Space then isn't 'empty'. Not only is there no such thing as a 'perfect vacuum', devoid of condensed matter but also as Minkowski's *'everywhere there is substance'*. The 'coma' of a 'comet' is just the 'bow shock' of a meteor, condensed from the medium by its motion, 'lit up' ever more intensely with density of solar or nebula gas & plasma. Coma intensify with greater speed through the condensate, as do the astrophysical bow shocks of ALL bodies & systems in space. Figure 1 shows a typical stellar bow shock, here 'lit up' by the Orion Nebula gases. Planets have similar shocks within the heliosphere.

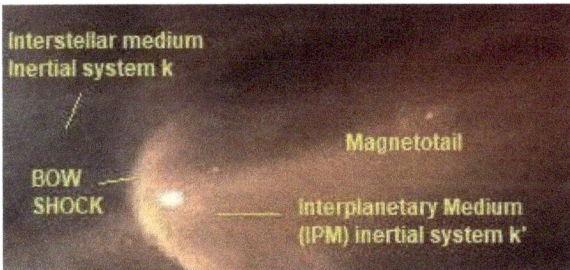

Figure 1. Star LL Orionis. HST Image. Bow Shocks & Magnetotails form the boundaries between inertial systems. NASA/ESE. Hubble Space Telescope.

Rationalizing impact debris from the 'DART' mission to Didymos, when free of *'confirmation bias',* also confirms the need for a local rest frame, so of a (Planckian scale or below) *fluid condensate* medium constituting all of space. (See Ch. 3. & 16. below). So no big *'dirty snowball'* fights if one reaches Earth!

Any Bodies 'Proper' (propagation) 'speed' in space then has the LOCAL 'datum' always required for it to be determined (A fact still missed or overlooked in much 'physics'). There is a *second* case of 'speed', only *apparent*, 'Co-ordinate', (**NOT *'Proper'*)** speed.

The *co-ordinate* speed case resolves the problem of two spacecraft propagating at near c doing near 2c with respect to the other, not *Proper* speed. English lacks a term for *'with respect to'* so I'll use '**WRT**'.

The nearest familiar concepts to the condensate are *'dark energy'*, or *'quantum foam'*. But the vacuum energy condensate was rather 'hijacked' in a swift 'smash & grab' raid to be applied as an explanation for the apparent *'accelerating expansion* of space' believed, or assumed, as demonstrated by *'cosmic redshift'*. The true cause of the cosmic (*wavelength* λ) redshift found inevitably emerges from the new physics, as due to the *expanding helical dynamic* of EM emissions & from the *local* datums for 'Proper' speed applying to orbital periods. Details emerge from coherent Alien physics. (See Pts 5, 6 & 10 etc.)

3. Not the old 'Aether'

The problem of the apparent *'action at a distance'* of electromagnetism (**EM**) & Faradays 'field lines', is also often ignored theoretically, but is solved by the new fluid medium. Faradays 'lines' are of common **spin *axis*** orientation, so electron spin polarity communicated via the *condensate* spin along with fluctuations in motion and density. The old Aether', was invoked to fulfil that function and to modulate EM speed ('of light') 'c', but hit unsurmountable problems as a fixed kinetic state when observers in ALL states of motion found light speed to be c!

The problem of reliance on maths over physical rationale emerges. Einstein gave up his physical 'heuristic' approach to adopt the *Lorentz* 'Gamma' *equation* allowing aether to be discarded. But he recognized *'space without aether is unthinkable'* [Leiden speech 1921] as it was only Aether's non-fluid "***immobility***" causing the problems. He wrote to Lorentz in Nov. 1919 [Kostro 1992] correcting his objection as; only the lack of a *single 'velocity'* for ether, *so "space has to be viewed as a carrier of physical qualities".* His 1924 paper clarified; "*ether should not merely be under-stood as imaginary.. something **real** in nature corresponds to it.*"

Stokes' 'dragged aether' had been even closer and recognised it's *'fluidity'* and many kinetic states. Fresnel agreed after Fizeau showed that light is (mainly) 'dragged' by flowing water. Unfortunately, Lodge *wrongly* used the *lab* rest frame rather than the **glass** frame in analysing apparent light paths in his spinning glass disc in 1893 raising doubts, with Lorentz in particular, about Stokes hypothesis, so physics theory went badly off track [see Jackson 2011]. McLaren saw and reported the error but was killed in the Great War before he gained attention. Stokes model wasn't complete, but the chance to disseminate and refine it was lost due to humans propensity for warlike behaviour.

The 'Sagnac effect' and ring gyroscopes proved kinetic 'drag' in new system IRF's after the initial speed *'extinction distance'* on crossing to the new systems kinetic state. But despite this logic and the required GPS Sagnac corrections, rationalisation in human physics *remains confused* (see 5 & 6 below).

The CORRECT mechanism is that the e+/- fermions propagated each side of shock shear planes and at surfaces of moving refractive planes *'couple with'* the EM (via absorption and re-emission) so dictate **new** *propagation speed c* on **both sides** of shear planes (both directions). Simple maths then gives these system transition zone (TZ) Doppler shifts.

13

Magneto-hydrodynamics (MHD) can well describe shock motions. In fact James Clerck Maxwell first found the plasma **TZ** and the different 'Near' and 'Far' field regimes each side of the TZ. He only just failed to see the *speed change* as the cause. Snell's Law of Refraction *also fails* at the TZ, as does normal Fresnel refraction, where anomalous 'Fraunhofer' refraction emerges. Both issues remain unresolved in human doctrine but also remain largely ignored! Both are coherently resolved below.

The TZ was characterised and modelled correctly as a *'two-fluid plasma'* (2 states of motion across shear planes) by Shumlak (2004) but is also overlooked or ignored. The plasma density dictates it's 'thickness' so the 'extinction distance' of the old *Proper* speed c as *new* speed 'c' takes over. J.D. Jackson's old 'Electrodynamics' text-book properly described the ('Ewald-Oseen) *'extinction distances'* in media of various densities. (High density = faster extinction).

Other lost opportunities to get understanding back on track included 1930 Nobel winner C. Raman's description of the birefringence during the TZ ***speed*** transition, and Dayton Miller's equivalent changing findings at increasing altitudes in the atmosphere on Mount Wilson. The changes are best explained as changing *birefringence* during the *extinction* (or 'modulation') of the old state and *'speed'* due to the progressive interactions, not as previously thought, & **not** complete ('*null*') on reaching Earth's surface.

To visualise the TZ speed change process it will be helpful to consider how the automatic gearbox of a car works. The speed input to a variably viscous fluid medium '*torque converter*' is progressively changed by interactions with fluid particles to a *new* output speed (a 'rate of rotation' in this example). Reynolds assigned his related 'numbers' to those transition rates. In the case of plasma in space, each particle converts the linear relative '**approach** speed' to a **new** speed dictated by the new common fermion spin speed via absorption & re-emission, commonly thought of as an '*atomic scattering*' process.

To summarise, the *old* 'aether' was erroneous. The medium is a FLUID, so has shear planes and vortex pairs between moving flows, as between all gas and fluids. The shocking new physics is then that; *LIGHT CHANGES SPEED CONSTANTLY to always *propagate,* at PROPER speed c *locally* in the fluid condensate. That's irrespective of local condensate 'motion' in its local background. All moving lenses then convert all EM propagation speed to *local* c/n before 'maths' processing by incorrectly applying the *time derivative* (so metaphysical) 'frequency'.

A 'golden rule' of nature is "ALL PHYSICS IS LOCAL". The familiar 'CSL' representing '*constant speed of light'* has been better defined, i.e. by Sirne & Bassi [2015] as *"continuous spontaneous localisation'* of light (EM) speed. This is the key to rationalising both SR & QM *'wave-function collapse'.* (See 20. Below).

Astute readers may now perceive how the data addressed by the Special Theory of Relativity is *physically* implemented. Einstein rationalised this with a new interpretation in his final, 1952, analysis, *'Relativity and the Problem of Space'.* But the change was seen as a major challenge to the 1905 interpretation embedded in doctrinal beliefs. It was also not formulated, so was poorly understood, and remains widely ignored. The corrected [1952] re-interpretation is rationalised further in sections 5 & 6 below. The 2nd key new truth to learn and embed is then the *mechanism* for localising c, which is that;

All electrons re-emit absorbed EM fluctuations at c in each electron's 'centre of mass' rest frame.

That concept will at first be unfamiliar but is not entirely original. It will be found as fully consistent with empirical data. Only once the full implications of the dynamics are grasped should the step of using algebra & (approximation by) maths models be taken. To help that stage, basic formulations are given in the 2012 fqXi essay *'Much Ado About Nothing'* (pages 9 & 17). See; Jackson [2012]. or; http://fqxi.org/community/forum/topic/1330

16

4. Refraction, Path Integrals and Lensing.

Before resolving the apparent paradoxes in SR an intuitive understanding of refraction should prove useful. Though quite familiar, Fresnel's refractive index 'n' for any medium can still only ever be determined by experimentation. The free surface 'fine structure' electrons are key to the process, and more so in the case of refractive planes 'in motion' through the local background medium, (see also 'Stellar Aberration' resolved for EM 'waves' below).

Existing valid terrestrial physics theory includes polarisation mode dispersion (PMD) in transparent media. Electron/positron interaction with 3D waves differentially *rotates the optical axis* of the EM fluctuations (including 'Light'). Where 'rotation' modulates wavelengths, the changes in periods are translated into, (so will be 'observed as'), variations in what we call 'colours'.

Plane wave incident angles lead to 'timing' delays in the re-emissions into the medium, so the matching fluctuations recombine at different angles to the refractive plane, defining the new axis found. This is akin to Feynman's 'path integral' derivation, which we derive intuitively as: Only the one of three lifeguards who chooses the fastest *combined angles* to first run across the sand then swim to the girl in distress gets there first, so saves her. She can then dismiss the others as they get there too late, so '*destructively* interfering' with each other!

'**Graded refractive index'** Lenses are revealing but seem not yet to have influenced understanding. In these 'GRIN' lenses the electron density varies, giving apparent so called '*curvature*' of '*light paths*'. Along with 'rays', both this '*curvature*' and '*light paths*' are *flawed and misleading concepts.*

More appropriate concepts and descriptions are an 'optical axis' and its 'rotation' at interactions. These will help advance understanding. We will discuss later the effect of spreading the particles of graded index lenses more diffusely and non-linearly, as found at densities which reduce radially around all massive bodies and systems in space.

Ex top NASA analyst Prof. E.H. Dowdye identified [see Dowdye 2007, 2012] the improved and physical consistency of such a rationale, based on refraction, including for 'lensing' by the varying density plasma halo's of galaxies and clusters. Increasing speeds as particle density reduces further from bodies also naturally produces the 'gravitational redshift' effect of EM wavelengths. (see also pt. 5 below.)

5. Einstein's Special Theory of Relativity.

I'll assume familiarity with 'physics' in the 1800's and issues with the 'Aether' posed by the speed of light being found the same by all moving observers. So, the word '*constant*' was used for light speed. But an option not considered at the time, or even now, was the simple option that *LIGHT CHANGES SPEED on ARRIVAL at 'observers', so on entering 'moving' media, **including moving lens refractive planes**.

When 'Luminiferous Aether' failed in logic, 'maths' was used to avoid the issue rather than searching for the logical process which Alien Physics shows is the correct solution; The interaction of EM waves with new 'inertial system' boundary shear plane TZ electrons localising speed to 'c' in each system. Maxwell's '*FAR*' field is then just the local background OTHER inertial system, *beyond* each local NEAR field. The simplicity of the truth is matched only by the confusion it seems to bring to those with older false beliefs embedded! (For 'inertial systems' read 'reference frames', **IRF**'s or inertial 'states').

Einstein's 1905 theory made a good fist of an explanation, but, forced to abandon his Heuristic views, it's complex 'interpretation' and maths held sway. Even Minkowski correctly said; '*everywhere there is substance*', and 'space' is "***infinitely many***

spaces in motion relatively". Einstein repeated that description in his 1952 're-interpretation' paper. He also noted that these moving 'spaces' had not before been *'thought of as bounded'*, and were *'logically unavoidable'*. He wrote "*bodies are not IN space but are 'spatially extended"* again as George Stokes *'Dragged Aether'*. Astrophysical shocks that form the TZ boundaries which interacted with and localised 'Proper' speed c, were later found (Fig.1).

The connection of astrophysical plasma shocks to Stokes or Einstein's ('52) model was missed. All dismissed or ignored such changes to the 1905 interpretation. Removal of paradox from the SR postulates also rationalises the interpretation of *'contraction'* & *'dilation'* and simplifies *'space-time'* to local sub-matter fluid condensate state k. NASA analyst E.H. Dowdye did find boundary TZ's using deep space radio telemetry data analysis. But, as NASA tries not to challenge doctrine, Dowdye had to leave to publish his *'Extinction Shift'* theory unaware of Einstein's matching 1952 re-conception.

These consistent models use the 'two-fluid' plasma of high coupling coefficient e+/- *fine structure* of TZ boundary shocks absorbing and re-emitting EM fluctuations. (See the constant alpha [α] Below). Einstein's '52 *updated* SR interpretation removes all apparent paradoxes plaguing the '05 original. *The Postulates now become entirely logical.*

The puzzle of astrophysical shock crossing analysis is also solved by applying the TZ *speed change* effect. NASA's *'Cluster'* probes revealed the particle kinetics, but the speed change across the shock was considered only as a step in *'energy'* during e+/- 'annihilation' over the Deby length. (This is better considered as vortex 'cancellation' and return to the condensate energy 'sink') All doctrinally based rationalization had failed. A problem with academic methods emerged when the correct hypothesis; considering the crossing as a **speed change** was indignantly dismissed by the head of a relevant faculty without study or analysis. It can't be over stressed that coherent alien physics shows *many* old and embedded beliefs need updating. To re-appraise:

1. **The laws of physics have the same form in all inertial reference frames**. Giving those *'frames'* physical substance with boundaries solves all issues. Within your car, bus, train or spaceship, whatever constant speed (kinetic state k) it has in the local background (datum) system state, all 'Laws of Physics' are then truly identical.

2. **Light propagates through empty space with a definite speed c independent of the speed of the observer (or source)**. 'Empty' means only free of condensed 'matter' states, and once we apply the hierarchy of bounded (fluid) kinetic systems to

'space' the logic emerges. Again, as has long anyway been known, it means; *"ALL PHYSICS IS LOCAL"*.

EM speed 'c' is really then a LOCAL PROPAGATION ('PROPER') speed limit within each of infinitely many (finite & bounded) *nested* inertial systems or 'reference frames' (IRF's). Any *apparent* speed, viewed by observers with other motions, is possible, so we have two distinct *cases* of 'speed', **PROPER** when measured at rest in the propagation medium, and **CO-ORDINATE**: 'measurable' *only* by rate of change of angle from any *other* state k', k" etc.

Emissions from Earth propagate at ~c in the rotating atmosphere's Earth Centred Inertial frame (ECRF) but *changes* to c' in the (orbiting only) Ionosphere's 'ECI' frame, then *also* to c" in the Solar Systems Barycentric Reference System ('Inertial Frame'). The change between each *local* case of c (c', c" etc.) is found in Doppler (wavelength lambda) shifts across each transition zone 'TZ' (commonly mis-derived as just '*energy*' shifts).

An error of 1905 SR was the assumption that there were any number of 'IRF's' at any position. In fact, only one state of motion at a time exists, but it may be as small as each moving electron. In this way Pauli '*exclusion*' can be interpreted as saying there can be only one '*state of motion*' at a time,

22

applicable to ALL inertial states, at all scales, not just those of electrons, which each form their own IRF.

Certain sets of twins may notice that they'll now stay the same 'age' wherever they travel. Only the Doppler shifted light signal *wavelengths* found on entering moving lenses makes far distant clocks **appear to run** faster or slower. But it's only the *wavelengths* that '*expand*' or '*contract*' to do so.

'Clocks' also cannot remain 'synchronised' when moved. Hafele & Keating's 'Round the planet' trip showed that all acceleration even changes atomic particle oscillation rates. They slow heading West and speed up on heading East, anomalous to most doctrine! In fact particles have a '*state of origin*', rate, slowing down when accelerated, including by gravity. Atomic particles (i.e. used as 'clocks') thus oscillate faster when in motion *against* Earth's rotation (westward) *and also* at higher altitude (less gravity), so closer to their '**State of Origin**' inertial (so *non-accelerating*) system kinetic state k. Again, this provides a mainly new, better, understanding.

As for 'space-time' Einstein said it; "*does not claim-existence in its own right..*" It's a Mathematical construct and needs replacement with the sub-matter scale 'fluid condensate' medium. Current confusion & anomalies are swept away, including the large pile 'under the carpet'! (see also Ch.17).

23

6. Trains & the Relativity of Simultaneity (RoS).

It may have been seen by now that Einstein's RoS, based on the 'constant' speed of light, wasn't fully comprehended or rationally resolved, in his '*Train Gedanken*'. The scenario & correct analysis follows; A train passes Andy (A) at speed v. On board is Ben (B), so we have two inertial systems (IRF's). Both A and B have oxygen masks so all can be in a vacuum! Simultaneous light pulse 'flashes' are emitted at the front & rear of the train, but *FOUR* light pulses, two *beside* plus two *AT* the ends of the train. A & B each see *their own two local* pulses simultaneously as each propagates at c in its *local* system IRF. They also see them at the same time as each other. But due to the propagation time A & B are no longer abreast of each other as B has moved WRT A, at v.

We already know that neither A or B can see the *others* light pulse as their lenses don't interact with them. However, let's assume A & B are 'LIT UP' by their respective pulses, so re-*emit light laterally*. Each one's re-emission is *only then*, visible to the other! As by then they're no longer abreast they may be fooled into thinking that each other's light pulses have propagated at speeds other than c! But 'Alien Physics' distinguishes 'PROPER' (propagation) speed ('c') from 'CO-ORDINATE' (apparent only or 'relative') speed, we've derived the principle of relativity (PoR) correctly, free of paradox. The train is just one of the *'**endlessly many** (finite, bounded)* ***spaces in motion within spaces'***, as Einstein stated in his corrected 1952 interpretation.

A more complete and intuitive understanding of the physical process may be gained in the case of **2** emissions, a short distance ahead of and behind the train. It is well understood, though often poorly considered, that as the light interacts with the windscreen glass refractive plane it *changes speed* to c/n glass (approximately 0.66c.) What human rationalisation has missed is that the speed c/n in the glass is **also** in & with respect to the new *moving* 'rest frame' of the glass (n). In reality that means that the speed CHANGE *IS **also*** by the trains v! So c'/n = c+/-v (+ see LT pt.12). Sorry for the algebra!

Adding v is a significant new correction, essential for a coherent understanding of *physically* rationalised physics. As the light passes from the glass into the train it's speed increases again to c (or c/n air), but now *'PROPER'* speed c in the (*'LOCAL'*) train rest frame. Light from the front screen does then have a shorter wave-length (Doppler blue shifted) than the light propagating along the track *outside* the train. The same occurs if the driver has the front screen open and the light pulse interacts with his lenses; it *then* propagates at c/n (lens) in the *lens* rest frame, so has 'changed speed' by c/n +v to conserve local c in **both** rest frames. The same happens at the rear window, the wavelength being increased due to the trains motion between waves, so is RED shifted as the light speed is changed to c in the new *train* rest frame. If the observer outside the train can track the progress of the pulses IN the train, he'll then find them at *apparent* c/n + v (+gamma) as they're 'carried' by the train, so with longer (redshifted) wavelengths. (Arbitrary *'co-ordinate'* not *'PROPER'* speed, is then an *'apparent'* speed only, entirely dependent on each observer's state of motion).

Any light NOT interacting with a window or lens as it passes through the train will have a far greater *'extinction distance'* (of the *old* speed c). If in a vacuum, it initially remains at the *old* speed c, so stay 'abreast of' the light pulse *outside* the train.

The 'extinction distance' in air is longer than glass but it reduces with density, which produces the altitude dependent data found by Dayton Miller in his varying altitude experiments on Mount Wilson. The Fresnel refractive index 'n' of the medium is a measure of the medium particle interaction delay effect. All refractive indices; 'n' depend on medium density, so dictate the 'extinction' distances, or the 'thickness' of the transition zone during which the birefringence occurs (due to **both** speeds existing during the progressive particle interaction process).

In the case of an open window and no refraction *into any new* train IRF, observer B in the train will 'see' the signal *before* observer A, because observer B will be closer to the source due to the motion of the train during the light propagation period. That unique case will **NOT** then give A,B simultaneity! The signals from front and rear will not then be observed simultaneously by B as the optical path length from the front will be shorter than from the rear due to his motion during the propagation time.

A fundamental new lesson to embed is that a light signal only propagates *where it is*, not **where it isn't**, So neither observer observes the other's signal, so cannot determine it's 'PROPER' propagation speed. As in much of man's physics there are more **cases** to consider than realised. A video analysing each case for Einstein's Train Gedanken and simultaneity is being produced and will be found on Youtube.

If not already familiar with Alien Science or fully comfortable with the simplified rationale of the derivations above, you are advised to re-read the whole section carefully to find the logic and gain that familiarity. Any such simpler concept will, if new, invariably initially *appear* to be 'wrong' and more complex, but due *only* to that unfamiliarity.

7. No 'Travelling Photons'.

The correct explanation of 'light' will again likely be shocking for most but careful thought will show it's veracity. The 3D 'wave' fluctuations propagate in the condensate medium with *any* polar axis angle, giving the helicity and ellipticity present in *all* EM energy including 'light'. The real reason for assuming that EM (light) consists of travelling 'photon particles' is that measurement interactions 'produce' quantized values, as, famously; *"the detector is part of the system"* (Bohr). But in 'photonics', as well as in QM, 'Measurements' are then the RE-quantizations produced BY, *and so only **manifested AT*** each interaction with cyclic particle motions before 're-dissipation'.

The 'detector' part of the system is initially the polariser electrons (as also in QM below) so the product is inevitably cyclic, 'pulsed' or 'quantized' by the detector particle orbital motion. The free fermions of the detector and lens 'surface fine structure' (electrons/positrons) are then *central* to 're-quantizing' energy and 'state on re-emission. *This is why quantized energy is all that can ever be 'found'*! But this doesn't imply 'travelling particles of light'. Only denying a medium allowed 'photons of energy' to be interpreted as ballistic 'particles'!

None of the apparently illogical and/or non-causal interpretations of experimental data are required once travelling ballistic 'photons' are replaced with local re-quantization. The so called *'quantum eraser time reversal'* effect, resolved by the wave model, should have been considered as proof of a flaw in starting assumptions, not proof of 'retro-causality'!

School children learn early that surface electrons congregate more at sharp corners. That includes *both edges* of the slits which light passes through, thereby producing the 'single slit interference' patterns found. If there are *two* slits, there are FOUR edges which re-quantise the light. All 'weird' effects can then be reproduced **causally** by the recombination of the 'split' light components, via constructive or destructive wave interference.

Huygens-Fresnel (H-F) wave-based theory was a valid model explaining why electrons have a high EM 'coupling constant'. H-F theory rationalises all experimental data and is the ruling paradigm of modern experimental and laser optics. H.A. Lorentz' dismissal of *'light quanta concentrated in space'* should have been allowed a far greater influence on theoretical doctrine.

So the shocking fact for many is that **Photon 'particles' must now gracefully retire** to their more minor role in photonics as the *'photons of energy'* from and AT re-quantization' interactions, which are jointly dependent on the amplitude, intensity, relative polarity, and degree of ellipticity of the fluctuations. This is a staple of Alien physics.

Updating beliefs in travelling '*photon*' particles will take some time, but only as would a new *word*! It seems to be advisable to commence with teaching the young, then wait a few decades! It is however seen as a very important change as it will lead to many of the inconsistencies in terrestrial theory being resolved in a coherent way.

That the concept of 'travelling' photons as 'light particles' is deeply embedded in the beliefs of most on Earth is just one of many examples of flawed assumptions becoming deeply held beliefs and so preventing advancement of wider understanding. Travelling particles of light are inconsistent and neither exist or are *required* to exist alongside the new *fluid* condensate 'baby' thrown out with the bathwater of the original immobile 'Luminiferous Aether.

Due to the human need generally to cling on to foundational doctrinal assumptions to underpin wider physics beliefs it may take inordinate time to advance wider understanding in this area. Much effort will be needed to replace beliefs with more coherent physics.

8. Spherically expanding wavefronts

Much of Schrödinger's work will be found as more valid. EM emissions expand from 'point sources' as spherical 'wavefronts' at local c. Any 'patch' on the expanding spherical surface has both 'transverse' and 'longitudinal' motions of the condensate medium. The wave energy dissipates in proportion to the spherical wavefront area expansion in space.

The wavefronts spherical shape can be distorted by moving regions of the plasma and gas encountered. Free fermion plasma has a (Fresnel) refractive index (n)=1 so is 'invisible', or '*dark*', except via its *kinetic* effects from relative 'speed' in the local background propagation rest frame.

Interactions with bound molecular gases are found spectroscopically. Both contribute to the '*surfaces last scattered*' concept of Scott & Smoots [2014] Nobel Prize analysis*, not yet rationalized into doctrine* but forming Einstein's '52 *nested, finite, & bounded* inertial systems. The dissipation of EM fluctuations in the medium increases the *radius of curvature* of EM helicity. It then needs large 'arrays' of multiple or very large dishes to distinguish. The logarithmically expanding helicity of Schrodinger's spherical expansion from sources has consistent foundations from Euler and in Fibonacci's spirals.

We also make the point here that 'optical axis' of these EM fluctuations is NOT always the same as the plane 'wave normal' (which is at 90° to the plane or surface). Optical axes are *rotated away* from plane normals due to *lateral motion* of the interacting particles. The 'rotation' modifies the ellipticity of polarity, described as 'ellipticised' Helicity. (Fig.2.) The small coil of charge orbits forming the larger coil is as now found in '*Hopfion rings*' of skyrmion 'strings', a valid area, but not elaborated on here.

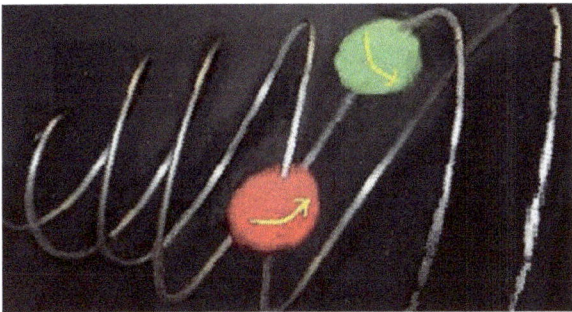

Figure 2. Ellipticised helicity of EM expansion. With or without twin helices and orbiting charges. PJ.

Ellipticity of helicity is a key geometry within Alien physics, which is founded on the 3D geometry of motion. The Helical ellipticity is widely applicable in topics from Cosmic Red shift to QM (see 20 below). Geometry as often applied in human physics can detract from rather than aid understanding!

9. Wave Particle Duality

A final word on light propagation is to consider the track of any 'point of charge' on the expanding wavefront as describing an expanding helix (of *any* ellipticity). Almost any 'view' of such a helix shows and entirely 'wavelike' progression. However, the view that counts in a 'measurement' is that of the axial 'interaction'. In this case the orbit of the helix is found to be *complete,* or 'closed', so is seen as a re-quantization, or 'particulate' rather than being 'continuous'.

So, as with all things, what you 'find' depends both on which way you look at it, and on your own relative vector, (Inertial system k) so explaining the *'observer dependence'* of any other than 'PROPER' (Propagation) speeds. The **changes of speed** at interactions with a moving lens explain apparent *'independence'* for the 'constant' proper speed c always found locally.

As indicated above, the so called *Quantum Eraser* interpretation shows a misunderstanding of the interaction process. The data are all explained causally above. Here the 'quanta' only manifest by and on requantization **AT** interactions (all slit edges and the back board). Only expanding 'wave' helicity in the condensate medium exists between the re-quantization's. Light then passes through both slits then recombines with either local constructive or destructive interference. Any asymmetry of glass crossing or reflection numbers between two paths can result in multiple full fringe differences! Such asymmetries can be identified in some familiar experimental set ups including those of Michelson and Morley, who did report unexplained 'fringe shifts'. An additional reflection alone, without an extra glass crossing, can delay EM transmission by ~0.7ns subject to wavelength, incident angle etc.

But Dayton Miller's findings suggesting a small residual 'aether drag' effect were correct. The non-zero effects showed that the speed of light entering Earth's rotating atmospheric ECRF from the ECI (ionospheric system) rest frame is not entirely 'extinguished' (Ewald-Oseen '*extinction*') before reaching ground level. The extinction rate varies widely with the angle through and optical depth of the atmosphere.

Miller's Mount Wilson data (see part 6) agreed with what is in reality (temporary) TZ '*bi-refringence*'. Babcock & Bergman's moving mirror 'reflection speed' problem was similarly resolved by the 'speed change' being **AT** Maxwell's Near/Far field Two-Fluid Plasma Transition Zone. [Jackson, Minkowski, 2012], and [Shumlak 2004].

10. True Cause of 'Cosmic Redshift'

Cosmic Redshift is the increase in '*wavelength*' of light detected from increasing distance, denoted 'z'. A simplistic assumption about 'cause' was made by the 2 groups competing to publish first on Type IIA Supernova etc. in their 1990's 'race' for a Nobel. The assumption the Doppler Lambda shifts were due to increasing recession seemed obvious, but it was flawed. We explain why, using **3D** oscillations.

When luminosity was found to reduce in proportion to the Doppler/Expansion solution no alternatives were seriously considered. Minimal consideration was also apparently given to the vastly smaller SIZE & MASS of galaxies in the early universe, so reducing their luminosity.

The 'Hubble Factor' has however, along with *different* redshifts for interacting galaxies, proved far from 'constant' and also lacks consistency in the assumption of accelerating expansion. 'Boundary' speeds greatly beyond 'c' were required, which confounded theory except for some rather 'arm waving' solutions. Edwin Hubble always believed, and wrote, that some **"unrecognized principle of nature"** caused Cosmic redshift. He was correct! Various alternative causes were proposed. Many invoke interactions with particles so creating the

'*Tired Light*' class of theories. (i.e. modified Newtonian 'MOND'). Particle interaction effects were quite well known, if poorly understood, but *none can produce the increase in wavelength* found when applying the 'constant' propagation speed.

The correct solution lies in the helical motions of EM energy, with an orbital path length which expands as a 'patch' on Schrodinger's spherical emission surface which expands radially at c with distance from source. This correct and consistent solution has not yet been considered in previous analyses, likely due to the poor understanding of the 'local' nature of the EM speed limit c. As the spherical 'wavefront' expands *radially* at c, the diameter, so also circumference of the helix increases. It will then take an increasing time to complete each orbital cycle, (including of any 'charge' involved) when constrained by the ***local propagation*** speed limit c.

Observers on the 'axes' of the helices will find the time for each (3D) wave cycle increased with time since the initial emission, which then manifests as increases of wavelengths, or redshift. This is an entirely new derivation of z, free of all past issues.

The particle interactions (plasma, gas & dust) then DO play a key role, modulating the redshift effect with constant re-quantizations in accordance with the Huygens/Fresnel theory, so producing the wide 'inconsistencies' found. The process also provides the (Scott & Smoot 2014) *'surfaces last scattered'* found across all of space and essential for fully consistent theoretical Astrophysics. The remaining 'anomalous' *blueshifts* found are caused by the motions of Earth, the Solar System and the emitters temporarily *'closing range'*, reducing separations.

It should be clarified that there IS an 'expansion' stage of the universe implicit in the corrected cyclic cosmological model. (see parts 21 & 22 below). Doppler Shifts of EM wavelength due to recession of emitters then DO have a limited role. Expansion (recession) is however only one short 'phase' of the full cycle, (as identified in 21 & 22).

A consequence of the opposite Chiral helicities also *propagating* in opposite directions (as AGN jets) is what is observed as an asymmetry of charge parity termed a 'CP violation'. To give a simple example: envisage a giant rotating bedspring on a flat-bed truck. If the truck is at rest the apparent 'motion' of the spring due to its rotation will be equal each side. But if the truck is in motion, the apparent spring motions are different, apparently violating charge conjugation 'mirror' symmetry or 'charge parity'. The same effect will exist both viewed from the side and, more relevantly, on interacting 'end on' due to the helicity Chiral handedness. (see also pt. 14.)

11. Stellar Aberration

While considering EM fluctuations, or 'light', the apparently intractable problem of deriving 'Stellar Aberration' (**SA**) with waves can be solved. Even now SA predictions are done 'by hand' (empirically) as, famously, *'no relativistic algorithm'* exists to calculate it. The reason is the false assumptions plaguing all 'Physics', and the answer, if not the mechanism, already exists. The aberration is NOT one of the very many proposed solutions but is due to a simple **rotation** *of the* '**optical axis**' of the wave fluctuations due to **lateral motion** of the interacting electrons at the kinetic Transition Zone ('TZ', see pt.3 above). The kinetic Sunyaev-Zeldovich (kSZ) effect found in space also results from this process.

The algorithms describing the effects are then those rather 'hidden away' i.e. within *'Mathpages 2.8'* described under; *˜Refraction at A Plane Boundary Between Moving Media"*. A physical explanation was also published in 2011 [Jackson, Nixey], but similarly the implications went un-noticed, so were not understood or disseminated. The 2001 IAU conference assumed that a relativistic algorithm for aberration *'would soon emerge'*. False assumptions prevented that. USNO Circ.72.2005 (pt.6) noted the problem was *still* unresolved. Only changing the old 'foundations' to those of Alien Physics, resolves the issues, consistently with Einstein's 1952 correction of SR. The Lateral motion of each particle being 'charged' gives an asymmetry of charge, *so **'rotates'** of the optical axis of re-emission at each particle.*

The Lorentz Transformation (LT) regime (see pt.12) is trivial at low (TZ crossing) speeds. We'll just consider the TZ between the (orbiting only) Earth Centred Inertial 'ECI' frame of the ionosphere and the (also *rotating*) inner Earth Centred Reference frame of the atmosphere the 'ECRF' ('WGS-84') inertial system. The transition between these two systems displaces a true star position to be seen 'ahead' of a systems motion, so is a 'reverse' refraction. Again, no newl maths is required The correct **physical** understanding is paramount.

We now explain what really happens at the dense electron TZ: Imagine one spherical electron of the trillions rotating with Earth, hit from above by the 'plane wave'. Due to the particles lateral motion the absorption of the wave energy in non-zero time is **asymmetric** to the plane normal. The electron, happily at rest among its companions has no idea they're 'moving'. It only experiences *asymmetric* absorption, *rotated away* from wave normals. The **axis of re-emission** is then also *rotated away* from the incident wave plane. The **optical axis** carries the information on apparent source position.

So wherever in the atmosphere, via a telescope or not, the apparent position of the star is '*ahead*' of it's true position relative to Earth's rotation (see Fig.3 below). The effect reduces, as does rotational 'speed', with latitude. The rotation of lights optical axis away from the plane normal is a new concept but is geometrically unavoidable. The long standing problem of Stellar Aberration for waves is intuitively resolved because a *whole plane wave* clearly **can't** 'rotate' on meeting the atmosphere!

Empirical proof of the kinetic reverse refraction of light in space plasma was found by the Very Long Baseline Array study of the dense plasma cloud moving into our AGN. (Pushkarev et al. 2013).

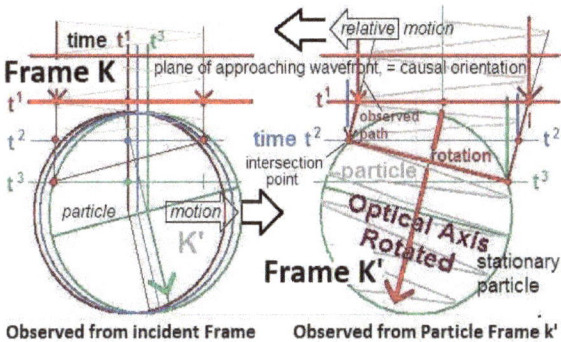

Observed from incident Frame   Observed from Particle Frame k'

Figure 3. Stellar Aberration. Waves charge laterally moving particles asymmetrically, rotating the re-emission optical axis. The apparent source positions are then seen to be *ahead* of the true positions. PJ.

12. The Lorentz Transformation & FSC Alpha.

The *'Lorentz factor'*, was invoked to help SR but in truth describes the increasing resistance to speed change through the TZ as electron density increases nearing relative speed c. Both the explanation and experimental data exist but the link to the LT has been missed. Free e+/- plasma column density near Earth is found commonly to be $<10^{14}/cm^{-3}$ and at $<10^{18}/cm^{-3}$, in the two-fluid TZ 'bow shock'. At near $<10^{22}/cm^{-3}$ plasma density reaches what's known in the specialism as *'Optical Breakdown'* (**OB**) Mode. Termed a *'dielectric breakdown'* phenomenon this slows then *stops transmission* of EM fluctuations as the two-fluid TZ plasma density increases non-linearly with relative speed. See also below.

LHC. In particle accelerators 'parasitic' electrons build-up in both bunch and pipe rest frames at high speeds and damage the pipe walls at near c. These were considered as problematic rather than a clue to new physics or an improved theoretical under-standing. Issues of conservation were raised, so they were renamed 'virtual' electrons. Not a 'solution' at all, that only 'hid' poor understanding. It went un-noticed that the power input and electron density curves when approaching c closely replicate the 'LT' curve. (see also p.45 below).

Increasing electron densities are in reality the main *cause* of the LT and the limit c. *They also provide the additional momentum found in collisions!*

LT. In Alien Physics; electrons & their density are the *primary* cause of the limit c and the LT. 'Optical Breakdown' (**OB**) mode is where the light can't penetrate when electrons can't oscillate when too tightly packed. Absorption and re-emission then fail nearing $10^{23}/cm^{-3}$. The Lorentz Gamma Factor is one case where maths modelling only approximates natures processes, and actually reveals nothing of the above process it models.

TZ. Maxwell's *transition zone* (**TZ**) is between his ***Near and Far fields***, all poorly understood concepts and with anomalies such as the failure of Snell's Law of Refraction in the 'far field' of the Fraunhofer refraction & 'radiation' regime. The TZ is the 'two-fluid' plasma vortex pairs 'photoionized', at rest to each side of the shear plane between kinetic states, The vortex pairs are as found at *all* fluid & gas shear planes. Beyond the 2-fluid TZ is a *different kinetic state*, (speed) so a different 'inertial system' or 'IRF'.

FTL. The speed limit c and Laws of Refraction are only valid in the LOCAL kinetic system, so not in the OTHER system, which is always the 'FAR' field. Any speeds tracked in that other system are only 'co-ordinate' speeds, so are only able to be found via

Pythagorean angular displacement rate, not by direct interaction, **so are not then limited to 'c'**. Distinguishing between the propagation 'PROPER' speed c and the 'CO-ORDINATE' speed (arbitrary apparent c+/-v) is then *far more important than can be imagined, and commonly thought.*

FSC, Fine Structure Constant; α ~1/137. The surface free fermion fine structure of matter at the kinetic transition zone (two fluid plasma TZ)   physically implements conversion of the EM wave *'approach speed'* (arbitrary c+/-v) to the *local PROPAGATION speed* c (or c/n) in the new medium rest frame, or 'kinetic state'. The Doppler shift of wavelength is created by the speed change. The **TZ** then regulates and localises the transmission speed c between co-moving media, via absorption and re-emission by the TZ particles on each side of the shear plane. Maxwell's Near/Far field TZ, thereby forms the 'domain boundary' zones in his 1862; 'EM' as the *'undulations of the medium'*.

The common assumption that the speed change into a new (i.e. lens) medium is simply to c/n, is wrong. The change is **also by the relative speed v of the lens in the background medium**, subject to the LT, as explained above. Clearly light in ALL lenses of equal index 'n' is identical irrespective of speed, providing irrefutable confirmation of the additional

'Galilean' speed change factor v between inertial systems, or 'kinetic states' k. (See Fig.4. below).

However, the 1/137 FSC has a better explanation than just implementing the kinetic transition which underlies SR. The fermion pairs condense at shear planes between kinetic *states in proportion to speed across the TZ* in the same way that vortex pairs condense at increasing scales in ALL gas and fluids each side of the shear planes within them.

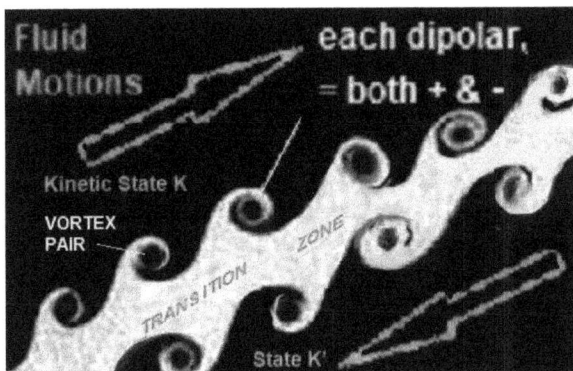

Figure 4. Shear Plane 'TZ' vortex pairs. Between fluid condensate motions, (as *ALL* shear planes). PJ.

The 'condensation' of matter particles from the smaller state into e +/- 'pairs, (be it from the 'Higgs condensate' or 'dark energy') is part of, or just one of, a wider **sequence of phase transitions**, the same ones seen as 'dimensions' or mathematical 'orders' in string/brane theories.

The sequence of pair scale radii 'steps' apparently continues to the limits of human perception both at smaller *and* larger scales. The pattern is equivalent to musical octave harmonics, as also found in the Chromatic Dispersion 'absorption bands' /steps in EM radiation.

In the condensed 'electron' regime the initial 'Higgs Boson', then Thompson, Compton and Bohr models scales or radii, are derived by different methods and differ widely, but with consistent 70MeV 'steps' between them. Mac-Gregor [2013 Ch.18] identified these phase steps as *'building blocks'* producing the smallest size of condensed matter; e +/- free fermion pairs. The experimentally well-defined symmetric e+/- energy 'ground states' get an "**α-boost**" energy factor of 137 times the particle *'radius'*, as *steps* in three 'production channels'.

$r_{boson} \times 137 = r_{Thomson} \times 137 = r_{Compton} \times 137 = r_{Bohr}$.

The 70 MeV **α-boost** 'building block' m_b is defined by MacGregors *boson equation* :

boson *(J=0): $m_e/ = m_b$* = 70.025MeV/c^2.

140Mev pions are created by the **boson-antiboson** equation: $E,\{r\} = 2e^2/r_{boson} = m_b c^2 + m_b c^2$ =140Mev.

As MacGregor found, the 'building blocks' in the three α-quantized particle production channels are, after the boson equation above;

Fermion (J =1/2) : $3m_e /2\alpha$ = m_j = 105.038MeV/c^2.

Guage boson (J =1/2) : m_{ud}/ α =m_{gb} = 43.17GeV/c^2

The energies are consistent with the fermionic Dark Matter hypothesis and derivation of gravity given herein, commonly considered to be in the tightest range of $10^{-3} - 10^7$ eV, also consistent with the high 'electron' (e+/-) densities (commonly ~10^{14}/cm^{-3}) found near Earth, as similarly found around other massive bodies and systems.

Mac Gregor showed that a 'solid' rotating sphere is an appropriate physical basis, but containing BOTH the electron and positron states, consistent with the dipole 'Majorana fermion' discussed herein (i.e. see section 14. below).

A physical meaning for alpha is thus identified, as a widely applicable gauge or 'scale' step between states, the full algorithmic basis of the parts of the sequence is presented and rationalised in Mac Gregors book '*The Enigmatic Electron*' (2013. see Ref.) so is not reproduced here.

13. Why is EM Propagation speed 'c'.

The astute reader may by now have seen where the limit c comes from. It's set locally by electron (or e+/- fermion) 'spin' speed. The absorption and re-emission of fluctuations by fermions is a complex subject. Few understand its full intricacies, including Alien Physics! The key factor is that the emission speed c relates solely to the *emitter* state, not the *relative 'approach speed'* of the waves. The basics can be simplified to an OAM value, or electron, 'spin speed' simply considered as an equatorial 'rotation' speed, dictating new local re-emission speed c.

There is commonly considered to be no problem with electron 'rotational' speed exceeding c, but we suggest the speed dictates re-emission speed. This is a valuable area for deeper study, but perhaps in future, once more fundamentals are understood and embedded. It will be essential to remember that the whole concept 'speed' is always **relative** to a local background **datum**.

Finally, it should be remembered that the validity of maximum **Proper** speed c is not only within each inertial system but also applies to each e+/- fermion. Physical refractive plane/plasma TZ's or 'two-fluid' *boundary shear planes* divide kinetic states, so *define* the concept 'LOCAL'.

14. Majorana Fermions and CP Violations

A firm hint about EM waves, electrons, positrons, & non-integer spins was given by Ettore Majorana [1937]. The Majorana fermion as '*its own anti-particle*' has become a foundational paradigm of experimental optical sciences but is still widely poorly understood theoretically. The coherent new explanation is simple: (See also 11 above).

Suspend old beliefs and re-consider fermion spin simplified to a spherical OAM rotation to gain an understanding. There should be no mystery in the 'dipole' nature of the spin (so Chirality of helicity). Encounter the Northern hemisphere and we find negative spin (anti-clockwise) or - 'curl', so it's called an 'electron'. But encounter a point in the *Southern* hemisphere and we find the opposite; *Clockwise* (+ curl) rotation. Now common electron 'spin flip' by around 180° allows all findings to be *reversed*!

This 'flip' can be achieved in several ways; 1) Moving around the particle & 'looking back from the other direction', 2) By use of 'refraction gratings', and 3) Magnetic field influences all common methods. Dipole rotations of physical bodies are inseparable, as the 'Ying & Yang' of older, wider understanding. We also often forget that *there's no 'UP' in space*. All galaxy rotation is equally clockwise and '*anti-*

clockwise', dependent only on how its observed or drawn. Planets *also* rotate in 'both' directions. We can only ever designate *above* & *below* arbitrarily.

So called 'Anomalous Magnetic Dipole Moments' are found when particles or helical EM waves also have linear displacement with respect to the observer. Chirality, or the 'handedness' of helical motions means that different states will be found when *observed* from opposing directions.

To visualise, consider the bedspring as discussed in pt.11 above floating at rest in space and rotating. The helical structure may appear to be moving *EITHER* to the left or right, subject to orientation. If also in lateral motion the **apparent** speed each way will be different due to the rotation. That effect applies axially so giving the +/- variation, which manifests as the anomalous 'CP violations', *not previously derived.*

For a sphere, consider how the tracks of a point on the surface vary in each case as it moves or as the observer, moves linearly past it. All detectors have electron 'polariser' fields with some orientation. The finding from the interaction will then depend on polar Chirality of the hemisphere interacting. Polarisation is discussed further under QM; Ch.20.

Note that repetition here is intentional, because all the physics is unified and also needs 'embedding'.

15. Non-Integer Spins

Seemingly less well known is that Ettore Majorana also derived not only 'non-integer spins' (also poorly understood theoretically), but also that these cases were *'infinitely variable'*. The physical explanation will help demystify the apparent 'weirdness' of QM. The completed solution and 'state' arises from **rotation on more than one axis concurrently**. These secondary rotations simply reproduce the 'non-integer' spins. In fact, the complete solution again involves the contribution of the spin axis angle of *polarisers* relative to the resultant states after interactions.

It is important to remember the 'signal' (or conjugate 'pair') polar axis angle is random and unknown. Trying to establish the angle in each case (which has all three degrees of freedom) involves *interaction*, which *changes* it, or 'destroys the information'. So called 'weak measurement' then doesn't resolve this problem. *Original* states can only be found for each pair if true 'single photon emissions' are used, and also after both (A, B) interactions, using relative setting angle data. The first step to comprehension is to visualise and physically understand the effects of *concurrent* secondary axis rotations.

For '**Spin 2**'; Take a 'point' anywhere on the surface of a rotating sphere. *Only when other axes are at rest does the point return to the observed starting position after one polar axis rotation.* If any other rotation on another axis occurs concurrently it will **not** return in one polar rotation. Viz. At half the speed on a 2nd axis, i.e. 'pole over pole'. After a full polar rotation, the original point will NOT then have returned to its starting point. It takes a ***2nd polar axis*** rotation to return it! The switch produces what's called a $720°$ *spin 2 'Spinor'*.

'**Spin Half**' is achieved where the polar rotation is only 180 degrees but is concurrent with rotation at the same speed on *either* orthogonal axis (i.e. a $180°$ *'pole over pole'* y or x axis rotation again). In that case our point will return to its starting position with respect to the observer in only ***half*** of a $360°$ polar rotation. Majorana's infinite 'number' or 'type' of polar spins can be all derived with just secondary rotations simply ***by varying relative rotation rates.***

Spheres are unique in having the ability to rotate concurrently on multiple axes without problems of momentum. The *product* of this measurement *interaction* produces the QM data. But even the new *'concurrent rotation'* analysis here does not extend to modelling the full causal measurement sequence we identify under 'QM' (see Part.20).

16. 'Messenger Particles' & Higgs Bosons.

It may have been noticed that the existence of a smaller scale 'condensate' medium, smaller than the smallest condensed matter, removes the need for so called *'Messenger particles'* moving in a 'void'. Media possess similar properties at each scale, including the *'frame drag'* effect of particles, found around all rotating bodies. This 'vacuum polarization' *direction* etc. is communicated 'at a distance' between adjacent massive particles via the smaller medium explaining Faraday's field lines of (nearly) *common polar angles* of field particles.

The Planck scale may be adequate for what's often called medium 'granularity', but the likes of Paul Dirac and Freeman Dyson correctly identified the apparent 'gauges' or 'phase transitions' revealing MORE than one gauge or 'step' down in scale. Wolfram elegantly derived a far smaller 10^{-93}m scale as the fundamental length, well below *'condensed'* phase. But now Haug [2023] Derives a 'Planck Fluid' direct from hydrostatic pressure, without knowing *'G'*, consistent with our SR 'condensate', quantum mechanics and 'gravity' as a 'density gradient'.

ESA's *'Integral'* probe CMB data suggest that any graininess must be over 10^{-48} and perhaps $\sim 10^{-75}$, over 57 orders smaller that the Planck scale. These are likely within the correct range and are again consistent with the higher orders of mathematical modelling, 'dimensions' or 'strings'. An analogy for the 'Higgs Boson', of very small but indeterminate 'size', can be found in the fluid medium forming the vortex pairs of e+/- fermions, (along with 'Quarks' where in a nucleus), which are the smallest form of *condensed* matter, coupling most strongly with EM fluctuations, yet remaining 'dark' as index n = 1.

The *'continuous or quantum'* argument is resolved because ALL gas or 'fluids' are made of particles, reducing *alternately* between 'granular' and 'fluid' views as scale or 'gauge' reduces. It should not be considered that a contradiction exists even below detectable scales. (Do also note that *analogies can* often be useful rationalising natures mechanisms, often scale invariant, but use with care.)

The 'Higgs Process' has been identified as an 'additional rotation'. That view corresponds to the e+/- vortex pairs, most commonly formed in the 'bow shocks' of condensate fluid shear planes from *'vacuum instabilities'* where disturbed. [see Behrle 2018]. Frank Wilczek's 2008 'Higgs Condensate' description of the medium is then valid. The 'Higgs-Boson' should then indeed represent condensate 'particles', but are commonly beyond man's ability to detect and measure with normal EM probing or spectrographic methods.

The existence of smaller states is clearly apparent from other 'unexplained' energetic effects including as those identified by Coulomb & Casimir. Imperial's electrical engineering head Prof. Eric Laithwaite demonstrated condensate energy effects in terms of gyroscopics and magnetism in various Royal Institution lectures (all archived on 'YouTube'). He demonstrated effects anomalous under physics doctrine but was largely ignored by the community, Alien physics fully rationalises Laithwaite's apparent 'anti-gravity', as a consequence of the condensate's physical effects of the same class as 'frame drag' on condensed vortex pair 'particles' and larger bodies.

The physics community response to Eric Laithwaite pointing out the effects he demonstrated had been '*missed*' by physics typifies the attitude which needs to be changed in mankind is to advance. Rather than address the matter as an opportunity to improve incomplete theory, complaints were made that Laithwaite was interfering beyond his domain!

In summary; ***Messenger 'particles' don't exist and aren't required***. *A fluid condensate medium free of 'Aether's problems transmits fluctuations*, and resolves wider anomalies in physics. See the next part. 'Gravity', and Ch.4 of; **The Origin of Gravity From First Principles.** Nova Pubs. [2021]. *'Sub Quantum Gravity: The Condensate Vortex Model'*.

17. Gravity

As well as carrying fluctuations, the condensate medium has radial distributions of pressure and density maintained by the vortices in the medium representing 'condensed matter'. The *'frame drag'* (of the Lens-Thirring effect and Gravity Probe B) of large body rotation is just one contributor to the pressure gradients formed. These radial gradients exist from the smallest scale upwards around each matter particle. The gradient simply *redistributes* the 'flat' ground state energy potential to the $1/r^2$ distribution around each vortex. Each particle then contributes to the total radial pressure distribution, *'additively'*, maintaining the macro force 'Gravity'.

Einstein's conceptual 'energy depression' is then a 2D version of the 3D pressure distribution, always requiring the 'medium' or 'field' we now identify. 'Brans-Dicke' scaler gravity theory, developed from Jordan's work, had credibility and veracity, but the high particle population in space which the model required wasn't found and confirmed until far more recently so was discarded, partly for lack of research funding in the USA for theories competing with SR.

Newtons $1/r^2$ *inverse square law* radial formulation was a good 1st order approximation near Earth but left anomalies plus much to be explained. Deep space probes need on-board artificial intelligence to take sightings and correct the predicted trajectories (Newton & GR). The corrections can be significant. Voyager 1 & 2 had many 'anomalous accelerations', all, like the 'Flyby' anomaly, unexplained by theory.

The distances from Earth now involved mean that significant time delays would occur if undertaken from the planet. 'Real time' course corrections by AI are needed to avoid increasing issues with greater distance. The oft repeated belief of many physicists that established theories are 'precise' and apply correctly to the whole solar system is quite wrong and NASA have the clear data confirming that.

A more accurate formulation is that completed by Bernoulli, based on the inviscid fluid flows around vortices. Vortex theory's inverse square law was apparently not fully considered as explaining Newtons maths. Newton lacked any physical, or now *'quantum gravity'* explanation. Newton did study the vortex gravity model, perhaps seeing it as competing, but he didn't try to falsify it.

Great care needs to be taken with visualization of how vortices produce gravity. The medium flow *around* a vortex is what **creates** the radial density/ pressure distribution. The 'gradient' created then produces the effect we call 'gravity', which the gas laws closely model. The orbital flow itself has no direct effect on either condensed fermion pairs or the motions of larger bodies. What DO impart a 'force' on 'bodies' are unequal opposite *pressure & density*, causing the bodies to 'gravitate' together in proportion to their mass and inertia. 'Equivalence' naturally emerges. Frame drag will cause variations over the poles (consider this effect as a prediction).

'Force' is poorly defined in mankind's physics, but a more advanced and consistent understanding will reveal connections between 'gravity' and the small-scale effects of 'nuclear particle' dynamics in the condensate. Orbital shocks & 'magnetotails' of Bow Shock plasma leave 'electron trails' on planetary orbital paths. The density variations and orbital motions mean the 'IPM' solar system rest frame medium within the heliosheath is not in just *one* consistent single state of motion (IRF):

NASA reports the solar system appears as *'layered, like an onion'*. Anomalous accelerations divide the spherical 'shell' layers. Layered motions do exist, as planetary orbits are a major influence on the kinetic states of the interplanetary medium. The significant 'Pioneer' & 'Voyager' anomalies are caused in the same way, but at the heliosheath 'shock' two-fluid shear plane TZ between the solar system & ISM rest frames, a 'flow' speed change of ~25km/sec. Binary star gravity has thus been confirmed as significantly deviating from predictions. (Kyu-Hyun Chae 2023).

57

There is then no need for 'graviton' particles, and the search for 'quantum gravity' has been focussed in the wrong direction. The vortex model of gravity has existed since W Thomson (Lord Kelvin) inspired by Helmholtz. The model has been explored by many since the 1870's including by Einstein, Tait and J.J. Thomson, (generating some 60 papers). Again; Vortex pairs 'pop up' at ALL medium disturbances (i.e. 'wing tip eddies'), which involve 'shear planes' between motions.

Bernoulli had expanded on Descartes & Huygens work, aided by his friend Euler and student Venturi. The full implications for gravity of Bernoulli's vortex analysis weren't revealed then or in later Vortex theories. Cyclonics, and also more recent work on Quantum Vortices, Spinors, Tori, Loop Quantum Gravity etc. all have a degree of related validity.

To provide a simple physical explanation; *When a medium is accelerated the distance between the particles increases, so pressure drops*. As medium speed slows, particle density increases, increasing pressure. The concept 'buoyancy' is also valid here. All cyclones have this *radial* pressure gradient (*independent of* *orbital* *vortex motion*) with the $1/r^2$ 3D spherical 'gradient' we call Gravity. Plan & 2D cross sections are shown in Figure 5 below. The close link between vortex gradients, nuclear Soliton profiles and Newtons $1/r^2$ formulation is clear.

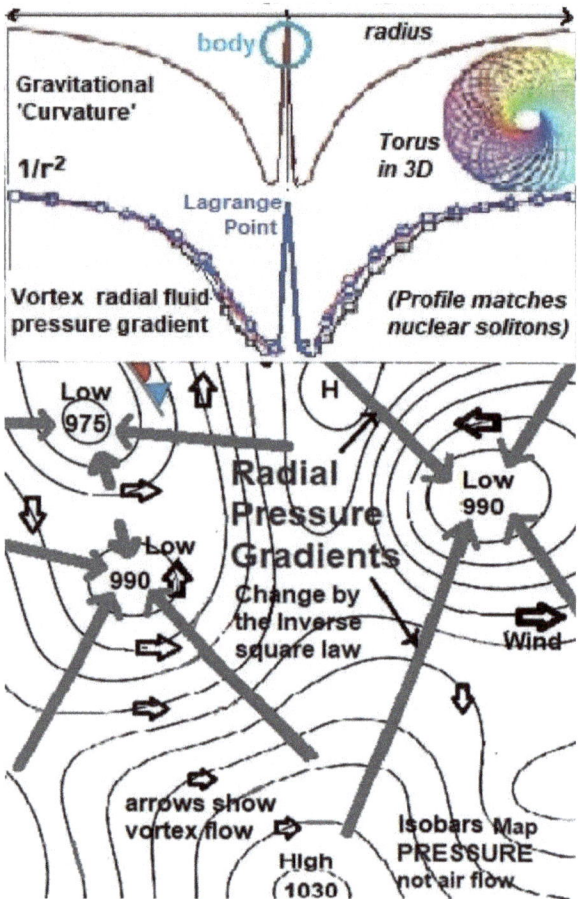

Figure 5. Vortex Medium Density/Pressure change with **radius, =** increased pressure *radially*, by $1/r^2$. In free space the toroidal (Chiral vortex pair) form reproduces all the 'anomalies' found.   P Jackson

Sailing yachts are really more *'sucked'* along, by this pressure differential. The accelerated air flow in constrictions between sails (the 'Venturi effect') reduces pressure. The lower pressure then 'sucks' the boat along with more force than the positive pressure of air slowed on the sails windward side. Both yacht sails and aeroplane wings then use the 'condensed particle' scale gravity. Relative medium accelerations produce **pressure/density** differentials and so the 'Bloch sphere' radial force vectors.

Misunderstanding or *lack of* understanding of the cause of gravity has led to the belief that 'gravity waves' exist *beyond* the propagation of the simple changes in measured potential caused by the spin motion of massive bodies. They do not!

The moon passing cyclically overhead Earth affects seawater via gravity cycles, so creates coastal and oceanic tidal flows These cycles apply in the same way that the motions of distant massive bodies produce the cyclic 'gravity wave' findings of the ('LIGO') Laser Interferometer Gravitational Wave Observatory. Separating the background noise interference from thousands of moving bodies and systems in the local universe is a massive task!

The Pioneer, Voyager & Flyby anomalies arise from medium shear planes & high vortex pair densities at planetary orbital radii (NASA's kinetically 'layered' structure), and frame drag 'tortional' anisotropies.

'Quantum Gravity' can then be simply explained, and gravitational effects reproduced in the radial density gradient of the fluid condensate medium around vortices. We know that ALL vortices, such as cyclones, create lower densities in the surrounding medium, with density and pressure increasing radially away from the vortex, so forming the radial pressure gradient. The overall spatial distribution complies with the 'conservation of energy' requirement, producing a 'flat' large-scale energy 'ground state' level between the local 'matter' peak and surrounding 'dip' at all massive bodies.

The problem apparently not resolved in historic vortex gravity models was that all vortices have an axis angle, so are spherically anisotropic, which does not reproduce the gravitational effects found around massive bodies. The same is true of toroids, which may be considered as coupled 'twin' vortices. The gravity of a body such as Earth and its toroidal ionosphere is largely isotropic. It is then similar at all positions, including the poles, perturbed only by frame drag, local oblateness and density variations. The solution to this 'correspondence' problem emerges from the random distribution of the vortex particle axes within any gravitating body or system. This randomness, along with the additivity of the gradients surrounding the particles, produces the familiar **spherical** pressure distribution or 'gradient' found as gravitation. The effect can be seen as the 'averaging' of the particle axis angle distributions within the body or system, giving the full spheroidal distribution found in gravity.

The anisotropies found in gravity naturally emerge, including the 'frame drag' from the larger body rotational axis, the increased potential over higher crust densities, the reduced local gravity found beneath massive structures such as the Great Pyramid, and also the gravitational effects of the dense e+/- condensed free plasma fermions of two-fluid astrophysical shocks and the less dense residual magnetotails. The latter would significantly contribute to NASA's 'onion' like layering of the IPM at planetary orbital radii.

The optical axis rotations of Einstein's 'lensing' are naturally implemented as refraction in gas & plasma halos, explaining why, as NASA's Prof E.H. Dowdye showed, lensing corresponds to the relevant halo-limits, not the gravitational gradient. Where two massive bodies orbit each other, the regular changes in potential at any distant point will be experienced as cyclic fluctuations or gravity 'waves'.

18. Dark Matter and Energy.

The origins of the data assigned as both Dark Matter and Dark Energy are identified above, but are summarised here for those reading only this section and, because the derivations are new so will have been unfamiliar. As a reminder: **'Dark Energy'** is the sub matter scale 'condensate' state, but *not* driving 'expansion' for the purposes of explaining cosmic redshift. The increasing redshift of distant bodies is derived above (Part 10) and also under 'cosmology' discussed below (Part 22). The scale is a significantly smaller 'step down' in size/energy from the smallest *condensed* matter, the fermion 'vortex' pair.

The dark *energy* condensate may be considered as having at least a Planckian scale 'granularity', but possibly, or perhaps with at least one further gauge step lower as suggested by Wolframs fundamental length of 10^{-93}m. (see Part 16 above).

Free fermions are the major component of the gravitational effects attributed to **'Dark Matter'**. Majorana Fermions (being their own 'antiparticle') are the electron/positron (e+/-) dipole vortices, propagated at high densities ($<10^{18}$/cm^{-3}) as two-fluid plasmas, at rest on each side of fluid medium shear planes. Astrophysical shocks form at the shear planes which represent Maxwell's Near/Far field

'Transition Zones'. These TZ's also form the inertial system *"boundaries"* defined in Einstein's (1952) *'logically unavoidable'* re-interpretation of Special Relativity. (also appendix (V) of the XVth edition of his book *'Relativity, the Special & General Theory'*).

His '52 SR re-interpretation repeated Minkowski's description of finite, bounded "infinitely many spaces in relative motion within spaces". As 'nested' IRF's. Anyone dismissing the conceptual paper as not implying important change for SR should note that he continued; "..***but is far from having played a considerable rôle even in scientific thought.***" It was new, correct, and needs to be understood and embedded as such, as we analyse here.

Excellent data exist on the extreme and chaotic magneto-hydrodynamic (MHD) turbulence of the shock transition zones bounding Einstein's *'spaces in motion'* in kinetic states k, k', k" etc. The particle dynamics are beyond mathematical accounting and even beyond 3D Lagrangian flow analysis. They are better rationalised with Chaos theory, as essential here as it's proved to be in atmospheric mixing and weather forecasting. Chaos theory has been rather too neglected by most of physics, indicative of the flawed popular desire to rely on Boolean binary or 'integer' mathematics instead of the more critical heuristic *physical* comprehension.

The free fermion 'Fresnel' refractive index is n = 1 which makes e+/- particles 'dark' to EM probing. 'Dust' and bound molecular gases are also now found at far higher densities than was expected and assumed to exist. These are both around massive bodies and within matter-based systems in space. Significant free electron densities have been known for well over 10 years. i.e. Schnitzeler [2012]

The (**non** - 'exotic') DM constituents identified here should now be properly recognised as significantly contributing to the total gravitational energy of a system. So, the non-zero gravity from particle spin, index n=1 (so 'dark' to EM) plus high *coupling co-efficient* means Alien Physics use of e+/- free fermion plasma as 'dark matter' means little or no contribution will be required from the various speculatively theorized new '*exotic*' particles to reproduce the 'anomalous' accelerations found.

See Haug [2023] below for the recent 'Plank Fluid' derivation of Alien Physics' fluid 'Higgs Condensate' medium replacing the old problematic immobile 'Aether'. It's found entirely from fluid mechanics. As also at;

https://www.scirp.org/journal/paperinformation?paperid=129922

19. Logic and Mathematical Modelling

We must mention the role of maths in physics. Abstraction and algebraic manipulation are useful tools. But it must be recognized as 'metaphysics', in that correct correspondence to the processes of nature is not guaranteed and often fails. Allowing 'time' quasi physical attributes as a 'dimension' can also mislead. Boolean, binary/'integer' maths can only ever *approximate* nature's non-linearities and hierarchy of gauges. The higher orders of maths addressed and modelled in String and 'Brane' theories are briefly referred to above. Such higher orders or 'scales' are equivalent to a reducing 'size' hierarchy of dynamic states and 'quantizations', equivalent to those of Set Theory.

The foundation of logic and popular mathematics, the *'Law of the Excluded Middle'* emerged from the 'indivisible atom' of Democritus. What man forget to do when taking the major step of *'splitting the atom'* was to review the validity of the indivisible atom. A consequence of humans ignoring the higher orders or smaller scale 'granularity' it remains true that, as Bertrand Russel put it; "*all logical systems are ultimately beset by paradox*'.

A new and better logic is required, using the firmly hierarchical structure of **Moda**l ('quantum') logic.

Maths will gain a more consistent structure. The closest existing structure is that of *Propositional Logic.* Here each **sub**-proposition must be resolved before a product can be applied in the potentially infinitely hierarchical 'next stage up' in resolving primary propositions. This structure is represented within mathematics by the arithmetic *'**Rules of Brackets'***, rules not always consistently applied throughout all of mathematics [Jackson 2015].

Equivalent with consistent 'modal' *propositional* logic; here each discrete bracketed function (x) represents a *'sub proposition'* to be resolved before it's *product* can then be applied as part of the; 'next sub- proposition up' in the 'nested hierarchy' of background states, closely equivalent to a fractal sequence. The hierarchical structure models nature in terms of Minkowski & Einstein's *'infinitely many spaces within spaces'* [1952]. These are the *nested* kinetic systems k or the 'flows' of equivalent local background states k', k" k''' etc. (see Part 5).

The 'local background IRF' hierarchy for assigning 'speed' starts from a single electron moving wrt others which may be part of a bulk flow. Each time the *proper speed* datum is *LOCAL*. A man walks in a train which is moving in the atmosphere's ECRF, which is rotating within the ECI frame /system which is orbiting in the Barycentric (Sun) system which is doing 16.5kps through the ISM, which is moving in the IGM, which moves in the Local Group, Cluster, Supercluster, Filament, etc. As *kinetically* nested 'Russian Dolls', each is defined by its motion & the boundary *surfaces 'last scattered'* localising propagation at c. LL Orionis, Fig.1. p10. has other 'tadpole' stellar bow shocks lit up by the gas Orion nebula gas. See also Pt.2. The rotating ECRF/ ECI Transition Zone is at around ~350km. altitude.

In both mathematics and logical systems, the key Alien physics lesson to apply is that the familiar '**Law of the Excluded middle**' underpinning binary and Boolean maths is a flawed and often misleading approximation. Consistent with Godel's theorems there is always a next higher order 'distribution' of states between 0 & 1. Such distributions are mostly sinusoidal curves, Bell curves or Gaussian/ Bayesian distributions. The more accurate '**Law of the *Reducing* Middle**' given above, is a *curve* between 0 & 1. As nature is 3D this is *helical*, but *plotted* as a sine curve between -1 & -1 with 0 as its axis.

Natures strong hints are often not recognised. Long distances EM signals sent in optical waveguides or fibres are effectively binary *'squared waves'*, not using the finer structure available. The signals tend to 'blend' and loose definition having to be 'squared up' again at relay stations. Binary signals sent into space by humankind then have limited ability to carry information and limited 'durability'.

Some concepts are all but 'invisible' to maths i.e. *'probability density'*. Found in the 1500's by Girolamo Cardano in his **'sample space'**. When Cardano's civilized medical methods were rejected by Milan's College of Physicians, he survived by playing cards and the 'Three Shell' game. He knew the odds, when removing one empty shell of the three, favoured a *change* to the initial selection. The initial *1 in 3* odds remained, even when reduced to two remaining shells [i.e. see Tijms 1978]

An extended argument arose between 'Parade's agony aunt Marilyn Savant and top mathematician Paul Erdos and the US Maths establishment over the 3 - choice 'Monty Hall' TV game show. Maths formalism said the odds were 1/2 for both options, 'changed' choice or not. But the maths convention was proved wrong, and Savant was proved correct! The invisible *'probability density'* concept became part of physics, but rather obscurely. Few broader lessons were learned or adopted by mathematics. Other such *'mathematical fallacies'* remain (a long list is given on *Wikipedia*).

Maths modelling isn't relied upon herein, for good reason. The intent is to provide the information for essential ontological understanding of nature and the universe required before numerical modelling. It's not an accident the major slow-down in the advancement of man's theoretical understanding coincided with the change to reliance on maths alone around 50 years ago. Algebraic formulations are needed to *verify* hypotheses but can't create them. Human brains can be very powerful as processors of information but are not presently utilised to their best advantage, constraining comprehension and so 'physics theory'.

Advancement needs a new 'kick start' which can't be provided by maths. Improving visualization and analytical skills in the training of Physics will be essential. Three-dimensional dynamic imaging is a key tool in Alien physics and needs to be better developed. Cartesian 'wire frames' can lead to only an imperfect understanding of natures' dynamic 3D 'solid' reality. The human brain is *capable* of more complete visualizations and rationalisations, but those skills now mainly exist only *outside* 'physics'.

It seems the re-introduction of those skills can be achieved, but difficulties will increase. The next section, about 'QM', should assist. The physical mechanism and causal rationale for QM will be presented. However, it will require more than **twice** the *'3-concept'* limit most humans can currently retain in mind from a single reading or lecture. Initially a 'check sheet' aide-memoir will likely be essential, then also repeated rehearsals to embed the new components, then to also reveal the new *implications*. Good luck, but read slowly!

20. Corrected 'Quantum Mechanics', & 'Time'.

A causal measurement sequence reproducing QM's predictions was published in a [Jackson, Minkowski 2022] Springer-Nature paper, agreeing with John Bell's view [1987] & 'theorem', so is now explained. Bells 'inequalities' go to the 'core' of QM, showing that the assumed 'conjugate pairs' with just one opposite spin pair 'superposed' can't re-produce the experimental data found without an unphysical or 'weird' process, not yet understood.

Bells' impeccable logic dictates that revealing which 'up/down' state manifests at Alice's polariser *must* dictate that Bob instantly finds the *opposite* state if he has the same polariser angle setting (or the *same* state if his angle setting is the opposite). QM then requires that if A or B *change* settings at the last instant, then that action dictates the **other** parties finding, thus implying instant '**action at a distance'**.

Bell showed that no so called '*hidden variables'* can re-produce those output relationship data found. Einstein's logical objection was that the implied instant action at a distance required 'super-luminal signalling' to explain the data, violating speed 'c'. QM assumed minimal 'starting assumptions', but there are always *hidden* assumptions. One was that particle spin as 'Orbital Angular Momentum' has only **one** momentum pair. Physics should be about hypothesizing and testing, so testing *all possible* assumptions to eliminate the inconsistent ones. Following this process leads to a *causal* QM rationale predicted by John Bell [1973 Ch.4].

We use Bell's own example of Dr Bertlmann's socks, always one red one green. The physical possibility **not** explored by QM was that the Doctor's socks were identical, **reversible** red with green linings or vice versa. *Bell's proof doesn't rule out that case.* We then simplify beam splitter 'pairs' to *identical* spherical 'particle' rotations. (helical plane wave momenta will also work) Studying the momenta shows each has *both* (+/-) polar 'curls' **plus** *linear* 'left/right' (or 'up/down') *'equatorial'* momenta, as in Maxwell's two inverse momentum pairs in OAM.

 On interaction with the (Alice and Bob's) *polarizer* electrons each one of these *two* pairs of inverse and orthogonal +/- states *exchange momentum*, with magnitudes dependent on the polar angles, so the 'angle of latitude', on BOTH spheres. (The tangent point here acts as proxy for the **absorption** angle). The outputs will then describe Wolfgang Pauli's *"classically non-describable two-valuedness"*, and also Paul Dirac's *'twin stacked inverse momentum pairs'* both classically and rationally (a first since Maxwell!) The *'rate of change'* of these 4 momenta between +/- 1 & 0 (max and min) across 90° of the spherical 'surface' is *non-linear* with the angle. The change rate is the same as that of the change of rotational speed of points on Earth's surface with Latitude. It equals the Cosine of the *latitude angle*; '*CosThetaLat*' (a constant for any sphere radius).

The *rates of change* of each curl/linear momenta are then inverse over 90°. Beyond 90° each value then *inverts*, to the *opposite* state (+1 to -1) at the full 180°. In the case of +/- '*curl*' the change at the equator is to the opposite (S/N) hemisphere. (South rotates clockwise {+}, North {-} is the opposite).

Each member of the two pairs of momenta can then be represented by a vector, so with direction and magnitude, incorporating the different 'curvature' at any point on the surface. Nothing on either the pair or polariser particles is then unphysical. Only the momenta *values* can be called 'superposed', but the vector values are 'physical', and so 'additive' on interactions. 'Measurement' is then, as long widely understood in theory, an '*exchange of momentum*', which can be determined by using vector additions *on each axis* (so giving a 'tensor' and new 'state'). Each conjugate pair of particles or emissions shares and maintains a polar axis angle (any and random on all three '*Block Sphere*' axes). The pairs then have 'anti-parallel' axes, so with the *opposite* hemisphere (+ or - polarity) always 'leading the way' from the source to the interactions. A and B's detector and polariser orientations can also vary in space through all three axes. The polar axis orientations then have the full three-dimensional freedom of a 3D 'Block Sphere' range of angles. Attempts at 'spin statistics theorems' have recognised a requirement to model momentum on all three (x,y,z) axes. What hasn't been recognised before now is the rotational 'curl' momentum set. Dirac's descriptions and graph of the two state pairs changing inversely over 90° *AND* 180° is correct as a 2D representation.

The change from vales +/-1 to 0 at 90° represents degrees of *ellipticity* of polarity. It should not be hard to visualise a sphere being rotated OFF its axis such that the ellipticity of the equator increases to become linear. If the pairs polarity is considered in terms of helicity the *ellipticity* of the helical path around the axis emerges as a valid physical concept.

Any point on a sphere surface (or oblate spheroid with an embedded torus) will describe an ellipse with major *and* minor elliptical axes subject to polar angle in relation to propagation direction. But here are *two* interaction stages at each (A & B) detector. The **2ⁿᵈ**, 'analyser' interaction has an important role:

Photomultipliers etc. After the polarisers modify the original states, the new states interact with the photomultiplier (Pm), photodiode, modulator or other 'analyser'. Ellipticity has a similar role at the aperture of the Pm, or at *each* aperture of a two-channel Photomultiplier, the ellipticity modulates the state entering the signal amplification 'cascade'.

Apertures are all circular as standard, so become increasingly elliptical, and so *smaller in area* with increasing incidence angle. **The area reduces to zero at 90°**. Note that instruments are 'tuneable' to record major axis amplitudes, so will find 50:50 at low ellipticities, as QM's 'predictions'.

A detectors twin Channels are set at relative 90°. When aligned to the signal axis, as one ellipticity goes to 0 the other changes inversely to become 1 (circular). The rates of change are inverse; ***by the Cosine of the incidence angle***. This 2nd 'ellipticising' process is what *'squares the modulus'* (explaining Born's rule) so producing the long thought of as 'mysterious' *Malus' Law* relationship; ***Cos²Theta***, consistently found experimentally, but *previously **unexplainable physically***.

Some will be most familiar with the 'magnetic field' version of the Stern-Gerlach experiment. Here the (electron etc.) particles are influenced by the polar axis orientation, set by the field 'magnetism', of the fluid condensate medium. Particle trajectories are 'deflected' by differing magnitudes subject to the particle polar angle relationship with the polariser 'field' angles, which are set randomly by Alice and Bob, including full 360 degree 'reversibility' of their setting dials, *so also of vector addition +/- values*. The new analysis based on complex vector additions will be found to have comprehensive validity.

A revealing aspect in this magnetic deflection case is that when any one (+ or -) state in the 'stream' undergoes the same process the **2nd** time they are similarly divided, subject to the setting ('field') angle. Logic confirms that both the (Majorana) 'dipole' view of pair particles and ellipticised wave helicity are valid, and that *polarizers **'re-polarise'** rather than simply **'measure'** particle states*!

The Three-filter' problem. An additional proof of the 'repolarization' effect of polarizers emerges in the solution to the case of a 3rd, 45^0, 'filter' slipped *between* two set at 90° to each other. All light was previously 'blocked' by the first 2 filters. Inserting the 3rd filter then 'releases' ~50% of light to pass through all three! But *'Filters'* is the wrong word to understand the screens, causing the confusion, lost causality and apparent 'weirdness' of the results. Each 'screen' is really a *'re-polariser'*.

The **'rotation'** of the light's polarity at and by the middle polarising screen changes the interaction at the 3rd screen from the original 90° (so blocking the full 180° components of the light) to block only a 45^0

component. The quantitative effects of the 45^0 rotations, as published in [2022] are as follows;

Cos 45^0 = 0.7071067, = transmission intensity ~50%.

So; (Cos $45°$ x Cos $45°$ = 0.7071067 x 0.7071067

= 0.49999).

The implications of the 'rotation' effect of the polarization axis of light by polarizers are wide and fundamental. However, these implications are *not usually applied* in terrestrial physics. This omission has giving rise to the apparent non-causality and long held misunderstanding of the results of the Quantum Mechanical 'Stern-Gerlach' etc. 'Beam Splitter' class of experiments.

Figure 6. below demonstrates the simple causal explanation of the 'mysterious' *three-filter* set up, Showing the solution, that polarisers are not 'filters but *re-polarize*. Note the 3rd polariser introduced set at a $45°$ angle between the initial two set at $90°$.

The new third 'filter' then *rotates* the polarisation (so, again, the lights *'optical axis'* is rotated) which effectively *'releases'* 50%, when previously it was 100% 'blocked'!

Figure 6. is from the [Jackson, Minkowski 2022], Springer-Nature *'Foundations of Physics'* paper *'The Measurement Problem, an Ontological Solution'*.

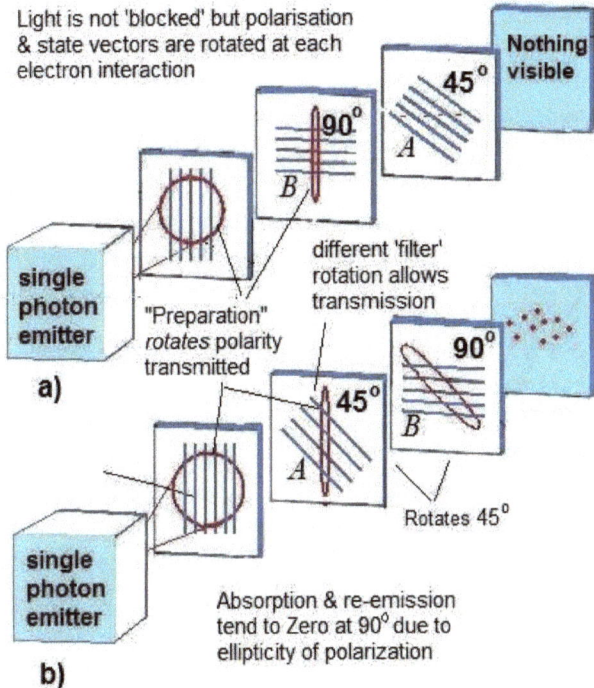

Light is not 'blocked' but polarisation & state vectors are rotated at each electron interaction

Nothing visible

45°

A

90°

B

different 'filter' rotation allows transmission

single photon emitter

"Preparation" *rotates* polarity transmitted

a)

45°

A

90°

B

Rotates 45°

single photon emitter

Absorption & re-emission tend to Zero at 90° due to ellipticity of polarization

b)

Three Filter Polarity Rotation. *Classical* **explanation of QM 'non-commutativity'** $H(A,B) = H(B,A)$

After Brukner & Zeilinger PRA 63 022113.2001. Fig. PJ.

Figure 6. Three-Filter Mystery resolved. Inserting a 3rd (45°) 'filter' between 2 at 90o 'releases' 50% of the light previously 100% 'blocked'. Showing polarizers REPOLARISE (so can 'spin flip') light. P.J.

What the sub-matter scale condensate medium can produce is shorter range 'tomographic' effects. In Tomography, strong magnetic fields can influence condensed matter particle orientations, often over significant distances. The range of these influences is subject to field strength plus other factors. So called 'quantum' effects can then be reproduced in a wider range of situations than the familiar 'beam splitter' experiments.

Bell's 'Theorem' Is hereby demonstrated as being 'circumvented' in the way Bell predicted in Ch.20 of his [1987] book " *'Six possible worlds of QM'*. Viz; *"the solution, invisible from the front, may be seen from the back.."* (p.194). The polarizer electrons are part of a large 'group' so 'stand their ground' and dominate 'pair state' vector additions at particle interactions. As is now clear, the vector additions can cause **reversals** of arriving particle states. i.e. +/- polarity. So A & B each change **their own** states.

Experiments have shown that the same reversal occurs when considered as helical Chirality of light. It only requires the right angle of reflection from a *'diffraction grating'* to **reverse the polarity** of light. As particle polarities are always 50:50, the *total distribution is the same*, but the assumptions used in the *'statistical analyses'* are wrong, leading to the obvious illogicality of QM, as identified by Einstein in the 1927 Solvay conference discussion with Neils

Bohr, and analysed later as the well-known 'EPR' paradox. Consistent with this analysis are 2 video's, showing some basics of Bell tests & 'entanglement'. https://www.youtube.com/watch?v=yOtsEgbg1-s and, https://www.youtube.com/watch?v=atqMU-VztQ4 These clearly identify the errors, and the tendency to use illogical beliefs over good science!

The Solvay and EPR argument should really have indicated that some error had been made. Bohr had strictly minimised his assumptions, but the *hidden* starting assumption was missing Maxwell's inverse *twin* momenta pairs ('curl' AND 'linear') in Orbital Angular Momentum, each reversing over 180° The 2nd momentum state was substituted by mystical & non-physical *quantum spin* resulting in QM's flawed 'rationalisation' of the experimental data.

Outcome: To clarify again; QM's 'conjugate pairs' had only *one* state pair and so the introduction of 'quantum spin' led to the '*logic*' that Alice or Bob must somehow change *each other's* finding '*at a distance*'. Experiments confirm that **if** this happens it would have to be instant! But, in reality, A & B *can each reverse **their own** states*, so **no 'action at a distance' is required!** The central 'mystery' of QM *is thereby resolved*. The effects of A or B flipping the spin (or 'polarity') of the particle meeting their polariser is that **NO effect on the OTHER particle state is required**. The simple consistency of the mathematics of this e+/- 'spin flip' case was confirmed by Sabine Hossenfelder's (March 2023) video following the full implications of the analysis published in 1922. (to confirm; Electron 'spin flip' and Chiral helicity flip are long established findings).

Compatibility with corrected SR. Sir Roger Penrose points out a fundamental contradiction between QM and SR arises from their treatment of 'time'. The *corrected* explanation of both theories here has no such inconsistency. In SR, time 'signals' simply have Doppler shifted wavelength, and so 'period' due to 'measurement' interactions at detectors with different kinetic state k ('speed' in the *local* medium datum). The speed changes are part of the repolarisation process as all emissions are at local c. (in the local 'near field' regime). Treating 'time' as a type of quasi-physical '*dimension*' for algebraic manipulation has mislead human understanding of nature. Other 'dimensions' are valid as larger and smaller 'scales', i.e. Smaller states the condensed fermion pairs. *Contraction and Dilation are **only** 'Doppler' effects on emitted signal wavelengths.* Either at the *emitters* TZ, or 'on arrival' if it's the *detector* that's moving in the local background.

'Constant' must be defined as *local* propagation or 'PROPER' speed. The LT is due to increasing plasma density. A brief 'check list' of correct interpretations follows, but first, two other matters are clarified:

81

Uncertainty arises from the inverse changes to momentum between 0 and 1 over the 90° quadrant of the sphere surface. It does **not** then represent the assumed *'position v momentum'* relationship but the momentum amplitude on **each axis** of measurement over the spheres four quadrants. 'Linear' momentum is then zero at the poles and maximum (1) at the equator. Rotation (or +/- 'Curl') is then 0 at the equator, and 1 at the poles giving maximum certainty of both polarities. Standing at a pole even aliens would fail to determine if they were 'moving' *left or right* (= 50:50 certainty). Wnen on the equator (so at 90°); *clockwise or anti-clockwise* **rotation** (or 'curl') is equally difficult to determine, so there is then also ~50% certainty for *that* quality. (**not** then *'position & momentum'*!)

Rotational anisotropy. When fully understood, the accurate re-formulation of this mechanism will result in a variation from the *'predictions of QM'*. It will also then however precisely **reproduce ALL experimental data**. The discrepancy between these two was identified by both Alain Aspect [1983] and Gregor Weihs [1998] in their ground- breaking 'time resolved pair' experiments. Aspect referred to it as an unexplained *'Rotational anisotropy'* the cause of which he couldn't identify. He had then to assume the issue arose from some system or equipment problem. Significant data had to be discarded to 'fit' QM's predictions. **All** his data is now explained.

Gregor Weihs et al, with Anton Zeilinger, employed different instrumentation, with an electro-optic analyser. But Weihs et al. had, and also properly reported, the same unexplained problems, so employed the same solution to recover a similar good fit to the 'predictions' of QM. This is a typical example of how confirmation bias can occur, and embed flawed beliefs. The physical mechanism of Alien Physics now identified reproduces & explains the FULL data sets of both experiments causally.

Note; A 2023 preprint by Annila and Wikström has confirmed an essential foundation of Alien physics, as titled; "*Quantum entanglement and classical correlation have the same form*". See;

https://www.mv.helsinki.fi/home/aannila/arto/Correlation.pdf

Other Corrected definitions include,

Conjugate pairs. Identical 'particles' moving in opposite directions, so with antiparallel polar axes.

Quantum spin. Replaced with Maxwell's inverse orthogonal momentum pair in OAM. So Polar rotation (+/-) AND linear 'equatorial' (L/R), changing inversely with each other over 90° by Cos-Theta-Latitude, both then *inverting (+/-)* over 180°.

Entanglement. The maintained anti-parallel 'Chiral' polar axis of particle or emission state 'pairs' of normal dipole states, so opposite hemispheres interacting with A & B polariser particles.

Wave function collapse. Is *modification* of states, i.e. polarity, speed and ellipticity by interaction.

Measurement. Repolarization of the orthogonal momentum pairs on exchange as vector additions on 3 spherical axes allowing local 'spin flip' of states.

Born's rule. Squared 'moduli' from two Cos Theta Latitude products due to 2 *subsequent* interactions, first at the polariser then the 'analyser'.

Retro-causality. A misapprehension due to using 'particles' of light not re-quantization. (see above).

Non-integer Spins. Rotations of *y or z* axis causing *'return to start'* in *any* 'number' of x axis rotations.

Probabilities. For any point on Earth's surface ask; How much is it; A) *Rotating* and B) *Orbiting* on a 0-1 scale. The first goes to 0 at the equator, the 2nd to 0 at the poles, giving Heisenberg's inverse relations.

ALL 'Interpretations' of Quantum Mechanics were misled by the hidden starting assumption missing Maxwell's inverse 2nd momentum pair in OAM. But of the 'Copenhagen interpretation' it can be said that no image of the moon is formed if no lens receives the emissions, *no image can then exist*. Changing the angle and motion of a lens that *does* exist will modify the image it forms, so giving a *subjective* reality. But with the correct causal mechanism, no '*interpretation*' of QM is required.

As John Bell showed and wrote in (Ch. 18 p.170) of *'Speakable and Unspeakable in Quantum Mechanics'*; [1987], "*..the founding Fathers Were wrong*" in precluding a causal explanation of QM experimental data. Few teach that fact, but it needs to become taught more widely along with the identification of the 'starting assumption' error and it's solution, identified here,

'Quantum computing' will be affected. No link may exist, but following the [Jackson 2022] paper both Chinese tech giants halted work. See also Pt.24.

21. Galaxy Evolutionary Sequence

As galaxies clearly rotate and evolve it's a source of bemusement that only the Hubble 'Tuning Fork' sequence exists. This was never intended to be a 'sequence', but only a morphological classification system. Galaxy evolution is an important subject affecting all inhabitants in the long term. What terrestrial theory misses is the 'cyclic' nature of evolutionary progression. Active Nuclei (AGN) and Quasar jets are at the core of the cycle. The correct evolution HAS been published, and also presented at the (2022) AAS conference, so it can be identified as the correct sequence. Ours is a typical 'barred spiral' galaxy, currently in mid cycle. We use this as a convenient starting point. But first it must be remembered that there is no '*up*' in space, only in gravity. *Humanity's 'top' view of the rotational direction of our galaxy is then entirely arbitrary!*

Barred spirals. A central dense 'bar' of average age stars is significantly visible in most spiral galaxies. Having no coherent explanation so far the bars vary in length and rotate on two axes, end over end with the galaxy and bodily on the long axis.. The bar gradually reduces in length and the speed of each end through the medium propagates the high star formation rates at the 'tips', from which the arms trail. The 'bodily' rotation around the *long* axis has one Chirality for half its length and the opposite for the other half. Those Chiralities have not yet been explained by terrestrial astronomy but prove the dynamics below. The two sets of 'arms' from each end also have residual long axis rotations Those Chiral rotations can only be produced by the peculiar dynamic processes of the bar and arm formation mechanism. (see below and Fig.7. p.88).

The Milky Way has now completed less than three rotations, but the Interstellar Medium itself also rotates, as a 'frame drag' effect. In the *local group rest frame* more rotations have been completed. Rotation speed also has 'steps' at virial radii. Contra-rotations of central *and* Halo regions are commonly detected. The rotational anisotropies are a natural consequence of each evolutionary cycle or of each 'iteration' of the galaxy having a new, often orthogonal, rotational axis.

All this time the AGN is growing by accretion, also producing outflows of matter, gas and plasma on the main toroidal axis. Terrestrial Astronomers have seen 14 whole stars ejected on the Milky Way disc axis in recent years. There have been more of these 'hypervelocity stars', and in *both* axial directions. As AGN power increases so does the ability to break down solar mass. From 2000-2001 'flares' have been found instead of hypervelocity stars, as stars have now started to be 'ripped apart' on entering one or other of two layered and opposing 'helicoil' paths forming the body of the torus (inside the disc, and all embedded in the oblate spheroidal halo).

Disc Galaxies. Open spiral galaxies evolve steadily into closed spirals and eventually into 'discs'. As rotations increase, the galaxy arms 'blend' slowly, transforming the morphology towards the common 'disc' form. The common designation of '*elliptical*' galaxy is a misunderstanding. Only the angle of view makes discs appear as elliptical. Disc galaxies have higher average ages of stellar populations, and normally no clearly discernible central 'bar'. Some galaxies can maintain the disc stage for extended periods. Stellar populations all age, giving the large 'red' disc galaxies often found. The majority steadily accrete main disc matter (from the inside first) to the central toroidal AGN, growing the AGN mass, rotation rate and gravitational power, so increasing the accretion rate.

Lenticulars. As the AGN accretion rate and mass both grow, a bulge starts to form at the disc centre, characteristic of a Lenticular or 'SO' type galaxy. The axial AGN outflows now contain more re-ionised plasma and have grown to the extent of forming the twin axial 'Fermi Bubbles' often found. The Milky Way itself already has such 'bubbles' formed by the opposing outflow 'jets', as they 'sweep around' the axis with their opposite Chiralities, as generated by the precession of the two flows as they meet and precess *around each other* at the very centre of the AGN torus. Figure 7 shows the full cyclic sequence.

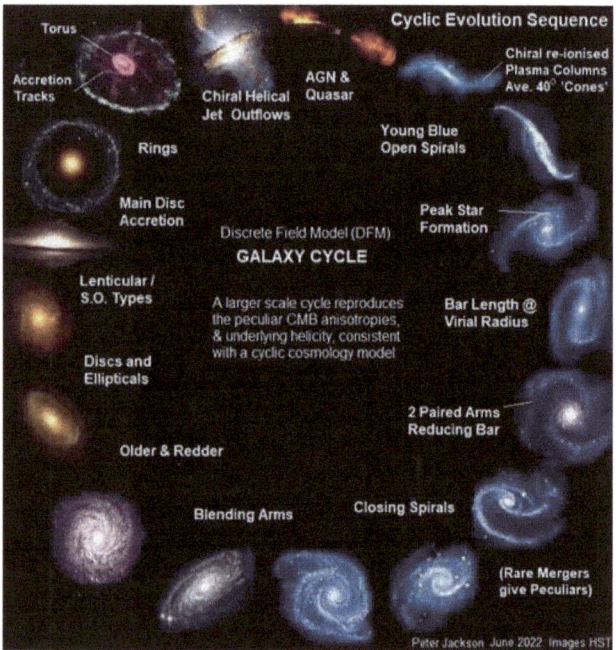

Figure 7. The **Cyclic Evolution model of Galaxies**, seen as a scale invariant process in Alien Physics. Peter Jackson. With NASA/ESA HST images.

Ring Galaxies. The efficient accretion of inner disc matter can leave the *ring* of outer disc material found at the Cartwheel Galaxy and 'Hoag's object', which both have high AGN masses. 'Trails' can be found from the ring to the AGN as the ring mass is accreted. In the case of nearby Centaurus-A the AGN 'Quasar' jet outflows reached full power long prior to full accretion of the outer disc ring matter. Such variations can reproduce the full range of galaxy morphologies including all of the many categorised as *'Peculiars'*. Jet outflows are often very focussed, many hundred thousand light years long, as, nearby M87, spread re-ionized matter in wide 'cones' into the local Intergalactic medium.

Quasars. It was first suggested by John Wheeler that the very powerful phenomena first termed 'Quasi-Stellar Objects' or '***Quasars***' must have some key fundamental purpose. The purpose has never been understood. Long confusion with so called 'Super-Massive Black Holes' (SMBH's) continues to confound man's understanding.

Quasars *are* central to galaxy evolution, in ways which can now be described. As Einstein identified [1939] and other such as Laura Mersini-Haughton confirmed [ref. 2014], 'singularities' the heart of 'Black Holes' do not and cannot exist. You *know* that all accreted matter ends up ejected again in the jet outflows so why does the misleading Black Hole concept remain such a firm belief when only high mass AGN *toroids* are found? Nothing can better demonstrate man's tendency to trust in old beliefs when data and logic reveal that the concept was flawed due to maths not resolving infinities. (see 'nuclear soliton' profiles page 95 & Fig.5. p.58.)

Increased mass from the accretion increases AGN rotation speed as ice skaters increase RPM by drawing in their arms. Increasing mass and rotation increase gravity so extends the effective accretion radius to the whole main disc. The accreted matter is accelerated ('heated') forming 'helicoil' wound paths around the tubular torus body with both Chiral layers. The friction between the two 'rips' the massive bodies apart. Lithium (Li6) will be found in the torus to aid this process as in Nuclear Tokamaks, explaining the anomalous paucity of Li6 elsewhere.

The 'Clifford Torus' well approximates the AGN's *'twin vortex'* toroid dynamics. The two dense hot flows eventually meet at the centre of the torus where they precess around each other. The hot matter is here finally reduced to mainly free fermion plasma when near full power. ***Each dense stream turns the other*** the final 45^0 required to power the axial jet outflows while imparting the opposing helical Chiralities found in the jets

As in any fluid, each jet outflow is 'collimated' into cylindrical layers, each outer layer slowing down due to friction and with the IGM. The shear planes between the kinetic layers generate new vortex pair production (as 'photoionization') so mixing the old matter with the new. The newly condensed matter increases total galaxy mass with each quasar event and new galaxy iteration (so coherently producing the anomalous galaxy '*mass growth function*').

Studies have shown the total mass in the IGM after the Quasar stage is greater than that of the original disc. The 'Merger' theory solution to mass growth is problematic as far too few mergers are found. New pair production at the shear planes is the real cause.

Each new 'pulse' from the jet root propagates at maximum speed c in the rest frame of the previous flow, giving the multiply collimated jet structure & explaining the *apparent* core pulse 'speeds' of up to 60c long found (also as the recent 'Mojave' survey) as 'co-ordinate' speeds. Local 'PROPER' speeds in each layer remain max. c. The *Rees-Sciama* 'closing angle' effect from jets initially faster than the light emitted also contributes towards the **apparent** superluminal speeds commonly found by rate of change of angle. (Geometrically via Pythagoras).

It may have been noted that there is no physical room for the Black Hole 'SMBH's theorised. Some characteristics of SMBH's do however result; The high concentrations of mass, the vortices leading to different areas of space and the red shifting of the receding jets *beyond* the visible wavelength range, so producing effective and real 'event horizons'.

The suggested 'spaghetti' formed on entering the AGN will also be cooked well beyond 'al dente' in the toroidal 'helicoils'. The opposing flows lead to what you've identified as the *'hottest place in the universe'*, at the 'cusp' where the two flows precess to 'turn each other' and create the Chiral outflow jet helicities, ubiquitous in outflow and later data.

When the old main disc 'fuel' is expended the AGN 'dies.' The re-ionized matter ends up with a long conical distribution around the old rotational axis. Cone opening angles can be near zero to over 45^0. (i.e. The 'Butterfly Nebula'). Contrary to common theoretical beliefs, but consistent with wide data, little 'feedback' occurs from the jet outflow matter back to the old disc.

The jet heads, partly of whole stars ejected early in the process, cool then 'clump'. The common 'ancient' stars of dwarf satellite galaxies are thus explained. Residual 'traces' of fast-moving gas and plasma as detected for example in the so called *'Magellanic Streams'* provide further powerful evidence of the past jet emission 'tracks'.

New Blue Open Spirals. The 'missing link' to the cycle should now be clear. The central section of the long 'cylinder' of jet matter slowly commences rotation to a virial radius on a new axis, forming the new 'bar' of a **new 'open spiral' galaxy.** The new rotation is initiated by the intrinsic & anisotropic CMB flow, so *'gyrokinetic rotation'* of bodies and systems. The flow is as found in the CBM dipole 'axis' and helicity, consistent with the cyclic model & further explained below. HST images of galaxies in this stage include ARP256, JO82345, & NGC1300.

As the 'cylinders' extending from each end of the bar follow the new rotation, each cylinder resolves to form a PAIR of trailing arms, so giving the 'twin pairs' of arms of all spiral galaxies. Evidence of the opposite helical rotational momentum of each half of the jet remain. Bodily rotation of the local arm is explained. The paired arms are both dragged *and* extended via the peak new star production rates found at the 'eroding' bar ends.

Stellar Scale Nuclei. The cyclic process and 'AGN' type dynamics can also be found at stellar scales. Notably at the 'Crab Nebula' core, is long described as a 'Neutron Star' as the outflow is the source of 'Gamma Ray Bursts' (GRB's). As the remnant of a recent, nearby supernova the gases not 'blown out' in the nebula have been accreted back to a spinning torus, producing outflows and GRB's in the same way as AGN. (see Figure 8). The stellar scale version suggests scale invariance of the process. The inverse nuclear soliton 'Mexican Hat' profiles (as shown in fig.5 p.59) show the accretion force inversion at the centre of the torus at that smaller scale. Note that an AGN's natural growth in power by accretion suggests a potential danger to man may result from correct nuclear Tokamak fusion dynamics!

Figure 8. Crab Nebula Core. Designated a *'Neutron Star'* but in reality a fast-Spinning toroid with axial outflows, as an AGN. NASA/ESA HST image.

Implications. Because all nearby civilizations will be affected by the cycle an overriding imperative exists to improve understanding if all are to survive in the long term. The sun has been recognised as a typical '4th generation' star, consistent with three previous iterations of the galaxy's stars. Now also explained.

The progressive 're-ionizations' of matter acquire a consistent explanation from this process, again all as identified. Cycle periodicy appears to have been increasing from the early universe in line with galaxy mass growth. At some 5 – 6 Bn years at the last iteration to the estimated 8 - 9 Bn years of this 'time around' iteration.

As well as the Mojave and other quasar surveys the Fermi probe found many thousands of emitters at high power, identified as 'black holes, indicating the population distribution over time. These data need careful re-consideration and re-appraisal.

The cycle time estimate coincides with a projected remaining life of the sun of some 4Bn. Years. The end of local solar system sustainability may not then relate to the next quasar event. There may be no hurry to relocate. Humankind has had a very short life in such cosmic terms but long term (two – three Billion year) targets should remain, be they travel to nearby or far distant systems.

The 50yr near 'halt' in the progress of theoretical understanding has been of concern to many. It is hoped that the more consistent 'new physics' we identify herein will help to facilitate a 'new start' to advancement of understanding in the corrected new directions.

The more consistent foundations given include the hierarchically 'nested' structure of the universe and the now evident 'cyclic' evolutionary progression which we'll discuss next.

22. Cyclic Cosmology

It has recently become ever more widely accepted that man's long popular cosmological models are somehow flawed. This is no surprise as little real evidence has been available to enable improved rationalisation. Instruments and probes are now advanced enough for data to reveal the flaws of old beliefs. It seems those beliefs are generally far too deeply embedded to allow change, thus this book.

Some have identified that, quite commonly, new theoretical 'patches' are applied to multiple older patches on patches. Old, flawed theories can be preserved, when fresh examination and analysis would lead to more consistent truths and allow understanding to advance.

A common rationale for dismissing change is that *'extraordinary new theories require extraordinary evidence'*. The basic Alien Physics herein is not at all 'extraordinary'. It will initially be found to look unfamiliar but on closer examination will be found to be demonstrably more consistent.

A cyclic model of galaxy evolution is remarkably consistent with many larger scale phenomena, so also with the scale invariant cyclic cosmological model of Alien physics.

Alien cosmology identifies the full evidence base to produce a new and more consistent model. An important basis of this advancement was the understanding of the *galaxy* evolution process, and its implications. In short, the galaxy recycling process and sequence described in the previous section is 'scale invariant', so repeats at all other scales. The universe is then a spatially finite system and evolves in a similar cycle to that of galaxies but at the far larger scale of 'universes'. There is then no reason to suggest that other universes cannot and don't exist in the same way as other galaxies do at the smaller scale.

Sound evidential proof remains beyond man's reach for now, but various cyclic cosmologies have been proposed. The basic cyclic cosmological model is correct, but most variants lack evidence and identify no *pre-* 'Big Bang / Crunch' conditions. A detailed fresh analysis of the characteristics and dynamics of the IGM following quasar outflow distributions in 'Fermi Bubbles' and beyond, reveals a full set of peculiar anisotropies consistent only with those found in the Cosmic Microwave Background (CMB) by COBE, WMAP and with more precision by the Planck probe.

Firstly: a 'flow axis' exists, with a dipole anisotropy, as would uniquely be found at the smaller scale off centre to the quasar jet axis. Even more unique is the 'anomalous' underlying large-scale helicity of the CMB radiation, which can only be derived from the smaller scale precession-based Chiral re-ionized Quasar jet outflows with opposing axial helicities.

The over-abundance of one (negative) Chirality or rotational 'handedness' also then acquires a cause. Opposing jet outflows have the *opposite* Chirality, giving the expected overall symmetry. All filaments are helical, from source precession. The finding of high z 'ancient' & large 'Methuselah' galaxies is also predicted by the cyclic evolution, as is the highly (re)-ionized constitution of the early-universe.

Methuselah galaxies can have one of two origins; 1. From the past iteration not, or not *yet*, accreted to be recycled, or 2. Earlier 'low power' outflow formations. The latter will be equivalent to the 'hypervelocity stars' in the axial outflows of AGN's while still building mass and power from the accretion, as the 14 examples recently found on the SagA* axis. In Alien physics the cyclic model is found to be by far the most consistent.

The Quasar stage duration within the *galaxy* cycles seems to vary widely. The Jets are commonly found many thousands of light years in length and up to hundreds of thousands. Jet lengths will be found to roughly correspond to galaxy disc radii, but over a wide range. For instance, NGC4889 in the Coma cluster is 1,300,000 light years (lyr) diameter against our mere 120,000 lyr. Our radius is then ~60,000 plus another 170,000 lyr. to the Magellanic clouds. Jet head progression would reduce to well below light speed, but the head would keep going for a long time. At say 0.1c average head speed the average jet duration at high power may be in the order of a few million years but with wide variation.

Scaling up to universe size may mean the duration of an outflow process replacing an instant 'Big Bang' could be nearer to a few billion years. The model reproducing the available data most consistently, is the scale-invariant 'large Quasar' Discrete Field Model Cyclic cosmological theory discussed herein. This model, reproducing all the peculiar CMB anisotropies, is then aligned closest to Alien physics.

Temporal *and* spatial 'multiverses' can logically then exist. The main consequence for humankind is that all or most constituent fundamental particles will have been a part of previous iterations of both the galaxy and the universe. 'Information' from previous states and perhaps past existences would then likely be retained, the extent dependent on AGN power and completeness of re-ionisation.

As a final note; Friedmann allowed an 'Oscillating' universe. Einstein also considered cyclic models in the 1920's rather than expansion. Both consistent.

23. Origins and fate of humankind

Our universe follows the cyclic pattern of galaxies, with no one 'Big Bang', and so will likely have had many iterations. It will, like galaxies, grow with each one due to new pair production, mainly at the collimated jet flow shear planes. Vortex pairs are produced in the condensate medium at the shear planes created by all fluid motion. A cyclic universe is then traceable to a distant 'origin'. We can say *'something moved'*, as a *'vacuum fluctuation'* to form the first single (Majorana dipole) e+/- fermion 'vortex pair' as the smallest scale of *condensed* matter. The additional shear planes formed by the vortex motion will then propagate further pairs, and 'there goes the neighbourhood'.

What could have 'moved' and so created the first *'vacuum fluctuation'* is entirely unknowable so at large. The question of the existence or not of an ultimate 'God' then remains unresolved. As Karl Popper pointed out *nothing* is really 'known', only guesses can be made. The aspiration of mankind should be to 'guess' more intelligently by using and analysing the full range of data. Relying on beliefs instead is the current widely prevalent method used by humanity, contradicting beliefs to the contrary. The fear of not having solid foundations to build theory from will be found unnecessary once the better grounded foundations provided herein are fully understood, tested, developed, allowed to prove their veracity and adopted.

The origin of life on Earth is quite well understood in the specialisms and so not a matter central to advancement of understanding. Along with details of Alien civilizations it is not then a topic to be discussed further here. But careful conditioning to Alien existence is important. You now have public and well documented observed evidence of advanced alien technology, and human responses, both on the planet and frequently also in space.

It is apparent that efforts have been made to reveal the existence of greater intelligences at the highest levels, including to political and United Nations leaders. At some stage direct contact may need to be made, but responses to observations so far and behaviour in general suggest humanity hasn't yet evolved an intellectual state adequate to join the greater galactic community. Secrecy, lies and active repression are gradually giving way to honesty and openness, but too many still deny the existence or even possibility of more intelligent beings. War, genocide, dishonesty, hatred, and the vast nuclear weapon stocks also suggest it's still too early for 'close encounter' direct contact events.

The panic caused in the USA by the Orson (& H.G.) Wells radio show seems unlikely to be repeated but further conditioning and intellectual evolution are still clearly required. The current concern is about the paucity of advancement of man's theoretical understanding since the start of the present *'shut up and calculate'* era. This era was born of poor ontological understanding but has led to a 'reverse outcome' in terms of comprehending nature's real processes physically. Maths can be an important modelling tool for approximating nature, but no more.

To avoid the real risk of 'extinction in the wild' man *must* advance. The advice and information on Alien physics herein should allow that fate to be avoided, if analysed, tested and adopted It seems that this book and process is all that can be done or should be needed to allow man to escape the present cul-de-sac' theoretical physics resides in.

24. Intelligent Life and AI

Replication of RNA in DNA reproduction is subject to the uncertainty element of the 'QM' mechanism as identified above. To simplify and visualise the process; picture the helical RNA 'string' as a long 'necklace' of spherical components, each with a particular orientation. Replicating the polarities of each element or 'key' of the string can mostly be done accurately as most momenta are certain. However, in cases at the poles and the rotational 'equator', uncertainty of one or other of the twin orthogonal momenta goes to maximum. There is then only a 50% chance of faithful replication of that, left/right (at the poles) or +/- (at the equator) momentum state. Small replication 'errors', or 'mutations' then naturally result.

Each new RNA string will then possess a small % of variations from the original string. Twins use RNA with the same coding, and also in rare cases the 'physical' keys can be replicated exactly, for identical twins. Where brain formation RNA closely matches, these twins may possess a form of 'entanglement' which manifests as thinking alike or even apparent as 'telepathy'. Human evolution may differ somewhat from aliens, but it seems the fundamentals are thought of as similar.

The mechanisms for *'intent'* & *'free will'* in human brains are ever close to being replicated in artificial intelligence. 'AI' is now recognised as potentially dangerous for organically evolved species. It will likely be a shock to some that these supposedly 'human' concepts can be closely mechanistically replicated with 'deep learning AI. Such replication doesn't infer 'pre-determinism' in causal process, but determinism itself is maintained within the limits of the uncertainties derived here.

The mechanism of 'intent' is simplified to impart a physical understanding. If sensory input suggests a 'choice' is required, the layered 'memory bank' and synapses produce responses by using 'routes' via 'switches' in each neuron. A 'choice' is then made leading to an action. If sensory feedback says that result was good, then chemical signals influence the 'memory bank'. The next time a similar choice is needed the same route and 'switch' setting will most likely be used. If the outcome was bad, different chemicals and switch settings are used, leading to a different response. The organisms then 'learn' from success or failure via 'feedback loops' from results. Organic processes are more complex of course, but it's been shown that that effect of 'free will' to make 'choices' can be re-produced 'bio-mechanistically' and likely outcomes learned. It may be an 'intent' to also try new, potentially 'worse' choices. Kant's *'Think for yourself'* and Hagel's

'Think differently' are essential human traits hard to recreate in AI without randomness. It may be inferred from the rationalised 'QM' identified above that low error rate 'quantum computing' is not achievable without using a 3D approach. AI relies on computers. A useful tool for complex and rapid computations, the potential dangers from AI must be understood and mitigated. AI will not quickly or reliably outperform & replace the best organic brains with 'better educated intuition' for the widest range of present tasks. But with better computing it will come, with those dangers.

AI is dangerous as a 'snapshot' *further embedding* flawed theory. Worse than Wikipedia, as most 'new' physics is precluded. Positive dissemination effects will then be outweighed by lack of advancement. Like computers, AI will also remain fallible and liable to error, both due to input flaws and mechanistic uncertainties. But human brains also require better training and organisation in their important *non*-mathematical capabilities to stay ahead of AI. It's suggested that AI will 'understand' what it produces soon, in a similar way to humans, and not limited by Searle's 'Chinese room' constraints. There are no good reasons for such comprehension not to eventually emerge. It has also recently been found that human intelligence measured by IQ (in a USA study) has actually started to reduce! That finding should raise concerns.

25. Hedgehogs & Foxes

The need to improve the rate of man's intellectual evolution should be thought of as an imperative. The skill of being able to 'build a model' in your head, test it and find implications of actions, can't be overestimated. We note that 'Chat GPT' etc. can't yet understand QM in any logical terms, in which it's equivalent to most humans!

As AI develops the option of not incorporating the above causal QM solution should be considered. AI can presently only kill humans by accident, but human belief in flawed theories may continue to stop progress in theoretically understanding nature and the universe. In that case AI will soon become aware that the slowing of intellectual evolution will render humankind redundant, so likely irrelevant and unnecessary, leading inevitably to extinction.

Good data now exist from most phenomena. The problem is in *analysis and interpretation*, which is commonly misled by reliance on doctrinal beliefs as foundations of interpretation. Karl Popper correctly pointed out that most such foundations are only built on mud! Freeman Dyson is considered to be a genius by most but often also as a 'heretic' by many challenged older assumptions and identified that advancement of scientific theory is: *'not about finding new things but about finding new ways to look at familiar things'.*

Marcel Proust had said similarly "*The real voyage of discovery consists not in seeking new landscapes but in having new eyes.*"

Also similarly Bronowski's "*All science is the search for unity in hidden likenesses*" is an important truth often ignored which must be central to teaching. 'Likeness' s an important word as all entities in the universe differ at any instant, not just, as famously, all snowflakes and grains of sand. The human concepts of 'identity' and 'equals' (=) will then require fundamental re-appraisal to help advance understanding.

Dyson usefully identified theoretical physicists as behaving invariably like '*Hedgehogs*' rather than '*Foxes*'. Hedgehogs use old paths and have very poor vision. When perceiving any challenge, they become defensive and prickly, curling up to a ball so are no longer able to see or advance.

Foxes on the other hand possess excellent vision, particularly when in the dark, and a wider field of view. Foxes search for, hunt down, investigate and fearlessly 'sniff out' answers. Dyson identified that Academia, and Physics in particular, requires, and needs to produce, far more 'foxes' if mankind's understanding is to start advancing again.

Only unhealthy scepticism about change or blind dismissals could then prevent progress, as could invalidly using old theories for falsifying new ones, which must cease, only then allowing progress!

Dyson also correctly pointed out that physics and astronomy are not only far from being coherently understood and rationalised but are inexhaustibly vast areas of study. None of the basic physics given here reduces the need for vast further exploration and study. Collaboration can be important as most individual human brains will struggle with the wider range of information required.

A desire to become the 'new Einstein' will not help lone researchers to overcome doctrinal resistance. Once change which should be found is that once physics starts to make the more coherent sense imparted herein, many more students should be encouraged into the discipline.

Older Physicists shocked by Alien Physics shouldn't despair. Once grasped you'll find the new physics simpler than old beliefs, and far simpler than it first appears. Knowledge of previous understanding will prove essential in deriving the changes needed and test current concepts. Much formulation work will also be helpful. Most Hedgehogs CAN now become foxes!

26. Climate Change.

Mankind is about to take the significant step of visiting other planets, yet care of the environment in which man evolved has been poor. All work to improve conservation of the resource is wise, but the greater picture should also be understood. The planet's climatic conditions and temperature levels have always evolved cyclically so including cooling, warming, floods and many prolonged droughts. There have been many cycles, including the five major 'ice ages' identified in the last 2Bn. years.

Reducing CO_2 in the atmosphere is one beneficial target, but the release of pollutants from ground currently frozen is unlikely to be much reduced and can be more significant. 'Change' is the natural state of nature and the universe and should be planned for. As Earth's climatic cycles can't be completely stopped by either human or Alien physics, detailed analysis of their effects is required. Dyson's surprising view that mankind can adapt to mitigate climate change is correct, but only when using a better understanding of nature from theoretical advancement.

Advice seems to be that greater emphasis should be given to longer term mitigation of the dangers to human and all other life. In addition to anticipating and responding to the consequences of the present warming stage, longer term thought should also be given to the next cooling stage, potentially more dangerous for humankind than warming. Earth's imminent polarity flip also presents dangers to technology not yet fully prepared for.

Life on Earth evolved to best suit Earth's particular environmental conditions, including Earth gravity, fundamentally affecting human physiology. Space travel increasingly reveals man's limited suitability for living in different environments and gravities. The improved understanding of gravity given above will not likely lead to solutions to the problems.

All planetary conditions differ. There can then be no assumption that space travel and populating other planets will prove a useful option for overcoming or escaping problems on the original planet. Home planets will always remain an invaluable and unique resource for the species that evolved there. Moving house should only be a very long term aim.

27. Non-invasive Alien Species

Humans may find it surprising if some rare animal species, under study by various groups, feared that humans may wish to 'invade' their territory. Aliens see that concept as similarly strange, but giving an interesting insight to humankind. It seems that no local alien civilization desires to 'invade' Earth. Here 'local' means able to regularly commute to Earth. It seems that various civilizations are able do so. All differ, some with more direct research methods, but none represent what humans would describe as aggressive or warlike. Tendencies of humankind towards such behaviour has been concerning but it's hoped may be reduced after this intervention.

Humankind is a valuable and interesting species with a wide range of possible future evolutionary paths. To influence such paths is a major decision, so here only general 'guidance' is given on which of the many already apparent and published paths should be taken to bring theoretical advancement back on track. The rest is up to mankind. There's an understanding that it's uncertain whether or not the guidance will be understood and followed.

Regarding methodology, A programme to passively influence human thinking has long been in place. Those with old theoretical doctrinal beliefs deeply embedded are least receptive. More open views or a better understanding of nature prove to help. Considerable effort seems to have been needed to comprehend what alien physics understands to be illogical beliefs, ideas and terminologies, 'alien' to Alien physics. Many theories not adopted into academic doctrine have some merit. Most of these are flawed or incomplete, but all should be properly studied and assessed.

Humans have a *'who* not *what'* culture, as well as selfishness and arrogance. Both must be removed from all research cultures. The Scientific Method ('SM') should work well but needs to be far better employed and applied. Old, flawed doctrine will then be less problematic to correct and advance.

Deceptions and errors also need to be exposed, i.e. in the radar range analysis of planets and the forced revisions to, or rejection of, papers that question old doctrine, for instance, challenges to QM, and the 1905 interpretation of SR. Using the SM consistently is essential to confirm the consistency and veracity of the Alien physics herein.

English is employed here as the most common Earth language, but its use and application varies widely in terrestrial physics, so apologies are offered to the many people not fully familiar with any terms used. The importance of humankind learning a common descriptive language to allow advancement can't be over-stressed. Other languages and cultures have *better* concepts in some areas of physics, but the evolution of the English language can be extended. Consciousness of subliminal communications either doesn't depend on any specific human language or the Aliens learnt English!

If the English language is to remain the primary written and spoken language of physics it will require to be developed & extended to allow better descriptions. New words are required, for example a word for "*with respect to*" (WRT) to simplify designating kinetic 'datums' in this kinetically hierarchical universe, in which no locally valid 'absolute' reference state exists.

Learning the true and basic meaning of +/- **'curl'**, and visualising bodies (such as planets) North pole (-) rotating opposite to a South pole (+) when viewed from 'above' seems beyond many. It should be one of the first basic skills taught in physics. Using '*clockwise*' and '*anti- clockwise*' for + / - Chirality is also unwieldy, + / - 'curl' is better.

Many Alien terms have no terrestrial equivalent and can't just be added to vocabularies. The new words must emerge from better physics.

Apologies are offered for the many and large specialist areas of nature and the universe not included herein. Other areas were covered, but confidence in understanding is so far incomplete.

Once the ways of thinking, methodologies and foundations are more soundly set any correctly constructed details should follow. But they will only follow if continued use of old 'beliefs' are avoided and cease to be the main methodology of science. The same must apply to mathematics once Kurt Godel's perceptive identification of the essential '*incompleteness*' of maths models is properly understood to be an 'absolute' limit.

From being considered as the '*language of physics*' maths must then be relegated to its correct correct role as a modelling tool for approximations, supporting rather than subjugating visualization, and mental rationalisation skills and capacity. The common fallacies, shortcomings and additional limitations of current formalisms must be corrected. The new mantra of physics should be; "*Understand first, formulate algebraically second*". Physics will then regain it's simplicity and beauty.

We must first comprehend and visualise physically, and only then use modelling 'approximations' with metaphysical symbols and numbers.

28. Improving human physics.

Beyond actual new physics lies man's methodology. Some have suggested man is in an epistemological crisis, principally emerging from the increasing inconsistencies now apparent in mankind's long-standing fundamental theories. Adherents to older doctrine will, rightly, continue to challenge the need for new physics and methods, but it's time for views to change to allow removal of those many *'patches on patches'* covering theoretical inconsistencies.

Human beliefs and methods have mostly emerged from a sequence of 'accident of history' influenced by social conditions. The advice herein identifies a wide but very basic range of shortcomings and the fundamental changes required. The summarised points that follow relate to theoretical physics in particular but may also have wider application.

1. Refrain from teaching theoretical doctrine and beliefs as *'facts'*, and widen curricula for theory courses.

2. New foundation courses need to re-teach 'how to think', critically, with broader overview, fearless honesty and refining 'pattern matching' skills.

3. Reduce monetarisation of scientific publishing; Generalize open access peer reviewed publishing with zero on-line or low publishing charges.

4. Create a word wide arXiv type archive with basic quality control & optional peer review level. Change Wikipedia from narrow doctrinal opinions with private censorship, or clearly state that's what it is.

5. Introduce a formal falsification grade system for proposed new physics. Do NOT judge against other theory however long established but employ the *Scientific Method* and consistency with data & logic.

6. Prioritise the gaining of physical comprehension of nature's causal mechanisms before reverting to modelling them mathematically. Recognise that *many puzzle solutions need no maths*, but that one clearly correct solution exists.

7. Root out and eradicate all types of dishonesty and bias. These include common 'confirmation bias', prejudices due to prior publications, poor analysis to comply with doctrine, and also hidden beliefs and assumptions.

8. Eliminate the fear of departures from the most popular doctrine for both educators and students.

9. Teach improved logic, falsification, visualization, research methods, history *as* history, and humility.

10. Change the culture of looking at 'who' rather than 'what' in assessing theory and papers. As A.E. stated; *"Unthinking respect for authority is the greatest enemy of truth."*

11. Remove the culture of entry to or retention within doctrinal theory as a 'Popularity Contest'.

12. Stop the predominant tendency to add new theoretical 'patches on patches' when new findings challenge established theories. New fundamental alternatives should be examined and analysed, so old inconsistent theory can be updated.

13. Teach more firmly that doctrinal theory and the current 'Laws' of physics are only provisional and need constant review.

14. Reconsider and remove false divisions between specialisms and disciplines to allow 'connectivity' between them and reduce use of inaccessible and gobbledygook descriptive language.

15. Teach that advancing man's physics is at least as much about finding hidden likenesses and seeing familiar things in new ways or with 'new eyes', as it is about new discoveries.

16. Improve language to enable simpler physical descriptions and intuitive abbreviations, i.e. for the relative; *'with respect to'* ('wrt') and 'clockwise' [S]/ 'anti-clockwise' [N] (+/- 'curl') etc.

17. Develop specialist skills formally into 'technical' and 'visionary'. Natural specialist skill areas should be taken advantage of, including some currently considered as on the 'autistic' spectrum.

18. Systematically ensure that the wide data and analysis of findings from the many advanced probes is made fully accessible to all for analysis.

19. Reconsider research funding policy to prioritise potential fundamental updates to ruling paradigms. Work on the minutia of current theory should not have priority. If establishment *'mafia'* re-emerge they should be publicly exposed.

20. Target grant funding for work on 'Identity' (i.e. no 'excluded middle'), Causal QM, and to examine inconsistencies rather than 'sweep them under the carpet'. Include teaching of 'Alien Physics'.

21. Divert funding from costly large scale projects into improving theoretical research & analysis and disseminating better physics. Mankind should be warned that once a nuclear Tokamak is correctly initiated it may grow uncontrollably by accretion, as in toroidal AGN and so-called 'Neutron Stars' such as the Crab Nebula core.

22. Give credit to those older educators who prove able to re-appraise their 'stock in trade knowledge' to achieve the above. Retire all who are not.

23. Be more rigorous with computer simulations to prevent confusion with reality. Similarly with AI, which relies on initial data input. (i.e. a flawed 'fact' embedded can result in nonsense emerging).

24. Encourage assumptions to be questioned and challenged, not dismiss a-priori. Test all hypotheses rigorously and constantly to try to falsify them.

25. Properly develop and enhance the human skill of *'pattern matching'* in a cognitive way to improve intuitive and coherent causal understanding. The move away from this area to relying only on 'maths' was a fundamental error which must be reversed.

26. Significantly reduce the costs of setting up, physically attending and participating in scientific conferences to promote wider inclusivity.

27. Use Alien Physics to develop a new series of more engaging and inspiring school and fresher physics texts to encourage the right kind of thinking, in the right directions, from an early age.

28. Re-focus judging of the 'predictive power' of theories away from precise quantification towards fundamental explanations for the many anomalies, unexplained findings and effects in current science and astronomy. It will be noted that Alien Physics resolves very many such unanswered questions.

29. Summary

The massive complexity of the universe prevents any true *pre*-determinism, but causality prevails. The true evolutionary patterns of nature in Alien theoretical physics will prove far more intuitive to understand than the confusing and disconnected beliefs of present terrestrial theory. It seems to be recognised that "*All theory is provisional*" but the consequences are not applied. Students should be taught not to 'fear' challenging established theory and to research for and reveal new physics. More widespread recognition is needed that many indeed *most* established 'foundations' are flawed and need replacement with more solid ones.

Methods used in teaching require fundamental change to allow better use of brain power, by the young. 'Foundation courses', as in other subjects, would allow students to be re-taught 'how to think' more freely and develop their visualization and analytical skills in a structured way. Teaching the *history* of human physics should be of the *historical* interest not indoctrination that old 'facts' are the reliable foundations on which to build. Intuition will re-emerge as a key tool once better founded. In short, a fundamentally new 'culture' is required to not only allow but also to *encourage* the revolution needed to replace the old methods and beliefs. That's the painful challenge facing mankind.

Those reluctant to relegate maths to a less primary role should study the existing problem-solving skills and proofs which don't require maths. Games and puzzles are useful examples. Chess masters don't require maths to think many moves ahead and win a game. the proof is in the *outcome*.

We can liken present Earth physics to a great pile of jigsaw puzzle pieces, all different and with very few consistently fitting together. *No maths is required to solve the puzzle.* The lack of a picture of the solution doesn't help, but that is given here. There can only be *one true solution* to the puzzle. The picture it forms may be very unfamiliar to most but it's correct, *needing no calculations to prove.* Only once completed can the correct formulations be developed to enable higher predictive powers.

A first step to better overview involves taking *'three steps backwards'* mentally for a broader view of the problems and parameters. Multi-disciplinary skills & understanding helps. Over-specialisation effectively creates 'blinkers'. Humans would also benefit greatly from memory training to broaden their 'database' & allow *'the new way of looking'* needed for better comprehension and problem solving.

Matrix mechanics was a useful approach but used false assumptions. Alien physics recommends a 3D approach to matrices for problem solving, needing practice and memory to use. *'Critical path network analyses'* of wide data input will then reveal correct solutions using far more 'joined up' thinking than common at present. The young have excellent skills at video, web and arcade games which should help if developed in the right directions.

The older have a greater propensity to 'run away screaming' from new physics and place *'heads in the sand'* to hide from or avoid it. It seems the fear of abandoning a *'stock of knowledge'* and having to learn afresh engenders fear in too many academics. Universities must however insist on fresh thinking, root out the troglodytes preventing advancement and remove them from the process to enable the required intellectual evolution to blossom.

The role of government as a funding source is also important. Research funding should be focused away from current 'naval gazing' and mathematics to developing the new advanced physics. Funding for those too bound by orthodoxy who refuse to look beyond old methods must be cut.

The vast funds required for, potentially dangerous, Nuclear Tokamaks and ever larger particle colliders will be far better spent improving human thinking skills. Far less cost is involved and the great risk of A.I. overtaking human intelligence is best delayed that way. Once familiar with coherent physics and better exercising and developing brains, the ability to receive subliminal suggestions should evolve. Further input from higher intelligences should then be possible, as needed if not 'on demand'.

The Alien physics imparted herein should not be thought of as a *'Theory of Everything'* (TOE) but will enable very significant progress once applied. Some key elements of *'Grand Unification'* of the correct elements of current theories do emerge.

At mankind's present intellectual evolution level, this book should be read carefully *at least* twice to help grasp and embed the fresh concepts The Universe should can only then be more readily comprehended. Unfamiliarity means much will at first seem 'wrong', and confusing, but a little persistence will reveal proofs its veracity and consistency.

Humankind seems able to have a great future, if it chooses to get advancement back on track. The 'checklist' below identifies some principle and fundamental elements of Alien Physics, to assist in the process:

A. Free e+/- Majorana fermions condense as twin vortex matter 'pairs' from a *sub*-matter scale fluid condensate. Likely Planck scale or less.

B. Fluid condensate motion produces *shear planes* and high 'vortex pair' production each side of two-fluid 'shock' transition zones. It also facilitates the *'action at a distance'*, needed since Faraday.

C. Pairs (e+/-) couple strongly with EM condensate fluctuations (via alpha), absorbing and re-emitting *ALL* Electromagnetic radiation, *so localizing physics*.

D. Re-emission is always at c in the particle centre-of-mass rest frame*, so **localising c in both systems***.

E. Free fermion (e+/-) *spin* speed dictates local EM Propagation speed c, as implied by $e=mc^2$ & alpha.

F. 'Constant' speed c is defined as LOCAL value (i.e. IN each ship or train) not with any 'universal' datum.

G. TWO distinct cases of *'speed'* exist: '**PROPER**' (propagation speed 'c'), a *LOCAL* concept, and '**CO-ORDINATE**' speed, (apparent, relative & arbitrary).

H. Emitters 'Near Field' c is in the emitter frame k. At the 2-fluid TZ it *changes,* to c in the 'Far Field' k'.

I. Lorentz's TZ physically occurs as the plasma nears *'Optical Breakdown'* (OB Mode) density, $10^{-23}/cm^{-3}$, increasingly reducing ability to oscillate to max at c.

J. Pairs are 'Majorana' e +/- 'dipoles', the lateral motion of which rotates optical axes, producing Stellar Aberration, the kSZ & related effects found.

K. Fermion Absorption 're-quantises' EM (inc. light), giving 'measurements'. Re-emissions then dissipate before the next interaction, as Huygens/Fresnel.

L. Special Relativity is as Einstein's '52 not the '05 interpretations, as a hierarchy of bounded spaces.

M. Spheres have polar 'curls' *and* (inverse over 90°) equatorial *linear* momenta, both + to - over 180°.

N. 'Uncertainty' of 'curl' is min. +/- at poles and of 'linear' momentum min. at the equator. Decrease of each is by CosTheta to **maximum** uncertainty at 90°.

O. Alice and Bob (A & B) can 'flip' pair particles, 180° with a 180° dial rotation, circumventing Bell's proof as he suggested, leading to rationalized 'QM'.

P. Polarizers and 'Analysers' both *repolarise* by the cosine of theta, producing Malus' Law. Cos^2Theta.

Q. All 'Non-Integer' particle spin is reproduced with concurrent rotations on polar **plus** *any* second axis.

R. Condensate 'frame drag' leads to Bernoulli's $1/r^2$ vortex *radial* pressure / density gradients, giving an asymmetry familiar to mankind as '*Gravity*'.

S. Galaxy evolution is cyclic, via AGN and Quasar re-ionization. The Universe shows a similar anisotropic dipolar, flow axis and helicity in the CMB.

T. Spiral galaxy bars & twin paired arms are formed by new orthogonal rotations (from flow anisotropy) of the jet matter columns out to a virial radius.

U. Mass Growth of galaxies over many eons arises from quasar jet 'pair production' not from mergers.

V. Gravity is the condensate density gradient from additive vortex motions, as Bernoulli vortex theory giving the $1/r^2$ 3D 'spherical curvature' of space.

W. Cosmic Redshift is due to expanding EM helicity, increasing orbital period, so apparent wavelength lambda, modulated by plasma zone scattering. CP violations result from Chiral +/- helical spin added to bodily motion of the spinning particles.

X. The concepts: 'Empty space', 'Neutron Stars' Travelling 'Photons', 'Gravitons', 'Big Bang', 'Black Holes', 'e+/- Annihilation', Wavefunction 'Collapse' etc. are replaced with coherent explanations.

Y. Atoms are divisible. A '*Reducing*' not '*Excluded* middle' (a Gaussian distribution or sine curve) gives the correct reality, logic and maths.

Z. 'Identity' is unique. All bodies in the universe are different: Galaxies, Planets, 'Candles', Snowflakes, Grains of Sand, Electrons/Positrons etc.

Key corrected descriptions provided herein include:

'Empty' Space, The 'Aether', 'Travelling' Photons, Helical waves, Gravitons, Gravity, The 'Big Bang', Neutron Stars, Black Holes, 'Worm Holes', Duality, Dark Matter, Dark Energy, Space-Time, Constancy of 'c', Coma, Bow Shocks, The 'LT', Uncertainty, Entanglement, 'Measurement', IRF's, 'Co-ordinate' Speed, Locality, Identity, Non-Integer & 'Quantum' Spins, TZ's, The 'Higgs Process', Superpositions' Cosmic Redshift, CP violations, SR, free e+/- fermion Space Plasma and the FSC 'Alpha'.

30. Epilogue

Neither *'translator'* or *'interpreter'* can adequately describe a writer's role when not involving use of any spoken alien language. Perhaps 'editor' or even 'ghost-writer' better describe the transcription job. As is the case in most of science, no conclusive evidence exists that this content came from beings of greater intelligence. This new physics appeared to have been received subliminally when all other sensory inputs are minimised, so in darkness and at the margins of sleeping. Sensitivity to subconscious communication seems to be needed. Some humans do seem sensitive, so potential for most others to do so with training should exist, opening a possible new avenue to advancing understanding.

It's likely that, without changes in thinking methods, closed minds & embedded beliefs would stop most recognizing new physics. It's crucial to change ways of thinking as current physics theory is badly flawed. An example of good visualization skill is ability to *see* the (+/-) physical rotations of spherical poles, rate of momentum change over $90°$, and implications. Possible sources of this physics other than Aliens may exist. The broad subject matter includes much in man's domain. But most is well beyond that domain so the possibility of genius can be ruled out.

The cyclic cosmology and similar galaxy 'recycling' sequence identified includes for both early and late accretion not leading to *full* 're-ionization of matter. The 14 'hypervelocity' stars found ejected axially at low AGN outflow power, suggest that planets might also be ejected. It is then quite possible that life may remain from previous iterations. Information from past civilizations may be conserved, held in brain cells. The concept of 'previous lives' *may* then not be as unscientific as most think. More advanced science knowledge may then also exist and re-emerge!

Ultimately each possible source of this advanced understanding can produce the same effects. Few entirely dismiss the possibility that more intelligent civilizations may exist or have existed, or that man could be able to progress in new directions. Our current methods, beliefs and physics are unlikely to be the best and most consistent possible.

If ANY time is right for a 'scientific revolution' or renaissance it should be now. Wider acceptance of the need for fundamental change has increased as growing inconsistencies emerge. Max Planck wrote that only when old physicists have died off can we advance. We also now need to produce far fewer 'cloned' students indoctrinated with the old beliefs.

Most of the new science, if not 'advice' herein has been tested by attempted falsification. *Apparent* inconsistencies have often proved to reveal *other* inconsistent assumptions, leading to one coherent unified picture. Some aspects have been published in papers and essays, often with much discussion ensuing. Some is understandably controversial, but no coherent or valid scientific falsification has been identified or sustained. Much of the content, was where cited, discovered and published by others, confirmed here as consistent with Alien Physics.

An initial short essay on most areas covered here was entered in the FQxI 2023 contest entitled *'Alien Physics, The Interview'*, under the topic; "*How could Science be different. And better*". The essay took the form of an Interview with the higher intelligence which is thought or assumed to be the source. Many will of course reject the new physics herein a-priori, so we finish with an apt quote from Freeman Dyson;

"The whole point of science is that most of it is uncertain. That's why science is exciting--because we don't know. Science is all about things we don't understand. The public.. imagines science is just a set of facts. But it's not. Science is a process of exploring, which is always partial. We explore, and we find out things that we understand. We find out things we thought we understood were wrong. That's how it makes progress."

References

Aspect, A, [1983] CH.20 Ph.D. Thesis, Trois tests expéri-mentaux des inégalités de Bell par mesure de corrélation de polarisation de photons. thèse d'Etat, Universite de Paris-Sud. p 265-267.

Behrle, A. et al. CH.16. [2018] Higgs mode in a strongly interacting fermionic superfluid, *Nature Phys*.

Bell, J.S., Ch.20. [1987]. ; 'Speakable and Unspeakable in Quantum Mechanics'. Camb.

Bernoulli, Jacob. Ch.17. [1683] De gravitate aetheris (in Latin). Amstelaedami: apud Henricum Wetstenium. Daniel Bernoulli 'Hydraodynamica' [1738].

Dowdye E.H. [2007] 'Extinction Shift'. Astronomische Nachrichten Volume 328, Issue 2.

Dowdye, E.H., [2012], Discourses & Mathematical Illustrations Pertaining to the Extinction Shift Principle Under the Electrodynamics of Galilean Transformations, Dr. Edward Henry Dowdye, Jr., 114 p.
http://go.glennborchardt.com/Dowdye12Extinctionbook

Einstein, A. *Relativity: The special and general theory,* App.V, (1952) Methuen & Co Ltd.

Einstein A. Ch.21. The Annals of Mathematics, 2nd Series, Vol40, No. 4 (Oct, 1939), pp. 922-936.

Einstein, A., (1924) "On the Ether." Schweizerische Naturforsckende Gesettschaft. Verhandlungen II. Wissenschaf-flicker Teil 105 (1924): 85-93**.**

Haug, E. Planck Scale Fluid Mechanics: Measuring the Planck Length from Fluid Mechanics Independent of G.

OJFD v13 No5.2023. DOI:10.4236/ojfd.2023.135019

Jackson. P. Nixey. R. Ch.11. [2011]. Inertial Frame Error Discovery Derives Stellar Aberration and Paradox Free SR. Ind.Acad.Edu.1921010 v1 2010. v9. 16.9.11. http://vixra.org/abs/1007.0022

Jackson P.A. [2012] Much ado about Nothing. fqXi 7th. http://fqxi.org/community/forum/topic/1330

Jackson. P. A., Minkowski. J. S. [2012] Resolution of Kantor and Babcock-Bergman Emission Theory Anomalies. Hadronic Journal. Issue No. 5. Vol. 35. pp.527-556. http://http://arxiv.org/abs/1307.7163

Jackson, P.A. Minkowski, J.S. [2013]. A Cyclic model of Galaxy Evolution, with Bars. HJ. Vol.36 No6. p.633-676. www.academia.edu/6655261/A_Cyclic_Model_of_Galaxy_Evolution_with_Bars

Jackson. P. Ch.19. [2015] FQXi Essay Contest; The Red/Green Sock Trick. Can Mathematics Demystify Nature? (Peer scored 1st). http://fqxi.org/community/forum/topic/2430

Jackson, P. A., Minkowski, J.S. Ch.20. [2022] The Measurement Problem, an Ontological Solution. Springer-Nature, *Foundations of Physics*, (2021) 51:77 https://www.researchgate.net/publication/352056822_The_Measurement_Problem_an_Ontological_Solution

Kaplan, H.G., [2005]. Ch.11. The IAU Resolutions on Astronomical Reference Systems, Time Scales, and Earth

Rotation Models. USNO. Circ. No. 179.
https://arxiv.org/abs/astro-ph/0602086

Kyu-Hyun Chae, [2023] Breakdown of the Newton–
Einstein Standard Gravity at Low Acceleration in Internal
Dynamics of Wide Binary Stars. The Astrophysical
Journal. 952(2):128.

Lodge, Sir Oliver. [1893] Ch.30. "Aberration Problems",
Phil. Trans. Roy. Soc. 184.

Mac Gregor, M.H. 'The Enigmatic Electron'. 2nd Edition.
El Mac books. Santa Cruz, CA. USA. 2013. ISBN 978-1-
886838-10-9.

Majorana E. [1937]. Ch.14. A symmetric theory of
electrons & positrons, I Nuovo Cimento,14 (1937),
p.171-84.

Mersini Haughton, [2014] CH.21. L. Backreaction of
Hawking Radiation on a Gravitationally Collapsing Star I:
Black Holes? PLB30496 Phys Lett B, 16 Sept.
arxiv.org/abs/arXiv:1406.1525

Minkowski, H., [1908]. "*at every place and time,
something perceptible exists. In order not to say either
matter or electricity, we shall simply use the word
substance for this something*". Gottingen Lecture.

Pushkarev A.B., et al., VLBA observations of a rare
multiple quasar imaging event caused by refraction in
the interstellar medium Astronomy and Astrophysics
555 (2013) A80. http://arxiv.org/abs/1305.6005

Routledge Critical Dictionary of the New Cosmology.
'LAST SCATTERING SURFACE'. (Def.) Routledge Inc., NY.